AF454462

Recent Advances in Antenna Design for 5G Heterogeneous Networks

Recent Advances in Antenna Design for 5G Heterogeneous Networks

Editors

Issa Tamer Elfergani
Raed A. Abd-Alhameed
Abubakar Sadiq Hussaini

MDPI • Basel • Beijing • Wuhan • Barcelona • Belgrade • Manchester • Tokyo • Cluj • Tianjin

Editors
Issa Tamer Elfergani
Instituto de Telecomunicações
Campus Universitário de Santiago
Portugal
School of Engineering and Informatics
Bradford University
UK

Raed A. Abd-Alhameed
University of Bradford
UK

Abubakar Sadiq Hussaini
American University of Nigeria
Nigeria

Editorial Office
MDPI
St. Alban-Anlage 66
4052 Basel, Switzerland

This is a reprint of articles from the Special Issue published online in the open access journal *Electronics* (ISSN 2079-9292) (available at: https://www.mdpi.com/journal/electronics/special_issues/RAAD5GHN_electronics).

For citation purposes, cite each article independently as indicated on the article page online and as indicated below:

LastName, A.A.; LastName, B.B.; LastName, C.C. Article Title. *Journal Name* **Year**, *Volume Number*, Page Range.

ISBN 978-3-0365-2964-6 (Hbk)
ISBN 978-3-0365-2965-3 (PDF)

Contents

About the Editors

Issa Tamer Elfergani (Member, IEEE, IET) obtained his M.Sc. and Ph.D. degrees in electrical and electronic engineering from the University of Bradford, UK, in 2008 and 2013, respectively. After his Ph.D., and since 2013, he has worked as a "Postdoctoral Researcher", and then as an "Investigador junior" at Instituto de Telecomunicações Aveiro, Portugal, within the Mobile Systems group. In 2014, Issa received a prestigious FCT fellowship for his postdoctoral research. The evaluation of the first triennium of this scholarship was given an excellent rating by the evaluators. He is now a senior researcher at the Instituto de Telecomunicações, Aveiro (Portugal), working with European-funded research projects, while leading technical activities in antenna design for ENIAC ARTEMOS (2011–2014), EUREKA BENEFIC (2014–2017), CORTIF (2014–2017), GREEN-T (2011–2014), VALUE (2016–2016), THINGS2DO (2014–2018), H2020-ITN-SECRET (2017–2020), POSITION-II(2018-2021), and Moore4Medical (2019–2023). Some of these projects have been successfully concluded, and some are still ongoing. He is currently working on (5GWAR) Novel 5G Millimetre-Wave Array Antennas for Future Mobile Handset Applications as Principle Investigator. Issa is an Honorary Visiting Research Fellow in Faculty of Engineering and Informatics, University of Bradford, UK. Issa is the Guest Editor for the Special Issue of *Electronics* "Recent Technical Developments in Energy Efficient 5G Mobile Cells"; the Special Issue on "Recent Advances in Engineering Systems Journal (ASTESJ)"; Guest Editor for the Special Issue of *Electronics* "Recent Advances in Antenna Design for 5G Heterogeneous Networks"; and Guest Editor for the Special Issue of *Electronics* "Unable RF Front-End Circuits for 5G and Beyond: Design, Challenges and Applications". Issa has around 160 high-impact publications in international conferences, journal papers, and books/book chapters, with a Google Scholar h-index of 20.

Raed A. Abd-Alhameed (Senior Member, IEEE) is currently a Professor of electromagnetic and radiofrequency engineering with the University of Bradford, UK. He is also the Leader of radiofrequency, propagation, sensor design, and signal processing in addition to leading the Communications Research Group for years within the School of Engineering and Informatics, University of Bradford. He has many years' research experience in the areas of radio frequency, signal processing, propagations, antennas, and electromagnetic computational techniques. He has published in over 800 academic journals and conference papers; in addition, he has co-authored six books and several book chapters, including seven patents. He is a principal investigator for several funded applications to EPSRCs and the leader of several successful knowledge Transfer Programmes, such as with Arris (previously known as Pace plc), Yorkshire Water plc, Harvard Engineering plc, IETG Ltd., Seven Technologies Group, Emkay Ltd., and Two World Ltd. He has also been a co-investigator in several funded research projects. He is the chair of several successful workshops on energy-efficient and reconfigurable transceivers: Approach toward Energy Conservation and CO_2 Reduction, which addresses the biggest challenges for the future wireless systems. He is a co-editor for the MDPI journal *Electronics* since June 2019; in addition, he was a Guest Editor of the *IET Science, Measurements and Technology* journal since 2009. He has been a Research Visitor of Wrexham University, Wales, since 2009. He is a fellow of the Institution of Engineering and Technology and a fellow of the Higher Education Academy and a Chartered Engineer.

Abubakar Sadiq Hussaini MSc, PhD, received his Post Graduate Diploma (PGD) in Electrical/Electronic Engineering from Bayero University, Kano, in 2005; received his MSc in Radio Frequency Communication Engineering from the University of Bradford (UK) in 2007; and received his PhD in Telecommunications Engineering from the University of Bradford, UK, in 2012. He is a member of IEEE, IET, ENF, and the Optical Society of America; has contributed to numerous publications; and is involved in European and CELTIC research projects. His research interests include radio frequency system design and high-performance rf-mems tunable filters, with specific emphasis on energy efficiency and linearity; renewable energy (wind, solar, hydropower, ocean energy, and geothermal), smart grids, batteries, electric vehicles, and power plants and equipment; robotics, nanotechnology, the Internet of Things (IoT), artificial Intelligence (AI), embedded systems, virtual reality, and augmented reality; radio frequency identification (RFID), software defined radio, cognitive radio (CR), radio resource management (RRM), mobile for development (mHealth or Mobile Health, mAgric or AgriTech, Mobile Money, and Mobile for Humanitarian Innovation), and optical communications. He has successfully attained several European projects, amongst which are the projects MOBILIA (2009–2011), ARTEMOS (2011–2014), CORTIF (2014–2016), THINGS2DO (2014–2018), and 5GWAR (2019–2021). He is a TPC member and a reviewer for many international conferences and journals. He is a guest editor for an *IET Science, Measurement & Technology* Special Issue, as well as a guest editor for a Special Issue of MDPI's *Electronics*—"Microwave and Wireless Communications".

electronics

MDPI

Editorial

Recent Advances in Antenna Design for 5G Heterogeneous Networks

Issa Elfergani [1,2,*], **Abubakar Sadiq Hussaini** [3], **Jonathan Rodriguez** [1] **and Raed A. Abd-Alhameed** [2]

[1] Mobile Systems Group, Instituto de Telecomunicações, 3810-193 Aveiro, Portugal; jonathan@av.it.pt
[2] School of Engineering and Informatics, University of Bradford, Bradford BD7 1DP, UK; R.A.A.Abd@bradford.ac.uk
[3] School of Engineering, American University of Nigeria, 98 Lamido Zubairu Way Yola Township Bypass, PMB 2250, Yola 640101, Nigeria; ash.hussaini@aun.edu.ng
* Correspondence: i.t.e.elfergani@av.it.pt

Citation: Elfergani, I.; Hussaini, A.S.; Rodriguez, J.; Abd-Alhameed, R.A. Recent Advances in Antenna Design for 5G Heterogeneous Networks. *Electronics* **2022**, *11*, 146. https://doi.org/10.3390/electronics11010146

Received: 17 December 2021
Accepted: 27 December 2021
Published: 4 January 2022

Publisher's Note: MDPI stays neutral with regard to jurisdictional claims in published maps and institutional affiliations.

Fifth-generation will support significantly faster mobile broadband speeds, low latency, and reliable communications, as well as enabling the full potential of the Internet of Things (IoT). This will open up the possibility for new services, such as tactile communications, smart manufacturing and cities, and enhanced broadband connectivity. Over the last years, industries and the academic community both have implemented tremendous efforts to bring these systems into the real world. In fact, the 5G system has been tested and deployed in several countries; however, this upgrade only exploits a small portion of the huge potential of 5G system. This is because the system still uses the sub-6 GHz band, i.e., the lower microwave frequency ranges between 3.1–3.55 and 3.7–4.2 GHz in the United States, 3.4–3.8 GHz in Europe, 3.3–3.6 and 4.8–5.0 GHz in China, and 3.5 GHz in South Korea [1,2]. Although some advancements have been accomplished (mostly in the software part and using existing 4G infrastructure) to improve the service, the transmission rates are still limited by the narrow bandwidth. In fact, when proposing 5G systems, both industries and academics agreed on the use of millimeter wave (mm-wave), i.e., with the operating frequency in the range of 24 GHz to 40 GHz [2,3] for significantly large bandwidth (large bandwidth is equivalent to large data rate). Unfortunately, the mm-wave has still not been deployed in the current 5G system. One of the main reasons is that the antenna, i.e., the end hardware part in the communication device that is used to transmit the electromagnetic signal, still does not fulfil the strict requirement of a 5G application.

The main topic articles in this issue highlight the recent advances in designing antenna systems for 5G heterogeneous networks. This Special Issue features 11 papers that present novel antenna designs synthesis along with effective approaches in order to improve the overall antenna performance. These 11 papers are concisely described as follows.

Yuxiang Tu et al. [4] give an overview and an introduction to microstrip filters, antennas, and filtering antennas (filtennas). Then, performance comparisons between the key and essential structures for these aspects are presented and discussed. Furthermore, a comparison between several RF reconfiguration techniques, current challenges, and future developments is presented and discussed in this review. Among several reconfigurable structures, the most efficient designs with the best attractive features are addressed and highlighted in this paper to improve the performance of RF and MW front-end systems.

Al-Yasir et al. [5] present a compact wide-band microstrip filter antenna design for 2.4 GHz ISM band and 4G wireless communications. The filter antenna has been designed, measured, and studied in three different dielectric substrate materials, which are Rogers RT5880, Rogers RO3003, and FR-4. The analysis was performed by using CST microwave studio software. A performance comparison for the designed filter antenna with different dielectric substrate materials and heights is presented and discussed using the same design configuration. The results obtained from each design indicate that the most suitable characteristics for a specific application can be achieved by using Rogers RT5880 dielectric substrate material. The structure is printed on a compact size of 0.32 λ0 × 0.30 λ0, where λ0 is the free-space wavelength at the center frequency. The designed filter antenna with the achieved performance can find different applications for a 2.4-GHz ISM band and 4G wireless communication.

Iqbal et al. [6] propose an isolation enhancement of two closely packed multiple-input multiple-output (MIMO) antenna systems using a modified U-shaped resonator. The modified U-shaped resonator is placed between two closely packed radiating elements resonating at 5.4 GHz with an edge to edge separation distance of 5.82 mm ($\lambda°/10$). Through careful adjustment of parametric modelling, the isolation level of −23 dB among the densely packed elements is achieved. The coupling behavior of the MIMO elements is analyzed by accurately designing the equivalent circuit model in each step. The antenna performance is realized in the presence and absence of decoupling structure, and the results show the negligible effects on the antenna performance apart from mutual coupling.

Dahri et al. [7] investigate a dual-resonance asymmetric-patch reflectarray antenna with a single layer for 5G communication at 26 GHz. The asymmetric patch is developed from a square patch by tilting its one vertical side by a carefully optimized inclination angle. A progressive phase range of 650° is acquired by embedding a circular ring slot in the ground plane of the proposed element for gain improvement. A 332-element, center-feed reflectarray is designed and tested, where its high cross polarization is suppressed by mirroring the orientation of asymmetric patches on its surface. The asymmetric-patch reflectarray offers a 3 dB gain bandwidth of 3 GHz, which is 4.6% wider than the square-patch reflectarray. A maximum measured gain of 24.4 dB was achieved with an additional feature of dual linear polarization. A simple design with a wide bandwidth and a high-gain asymmetric-patch reflectarray make it suitable for use in 5G communications at high frequencies.

Altaf et al. [8] study an antenna design for a UWB MIMO antenna system with an isolation of −20 dB. The UWB response is achieved by designing a stepped-shaped antenna while the isolation is improved by using slotted stubs etched in the ground plane within the radiating elements. The system is fabricated on an FR4 substrate. The circuit theory is used to model the circuit model and compared the results with the EM model. The simulated, measured, and the circuit model results are in good agreement. It is found that ECC and DG for the proposed system are 0.15 and 9.85 dBi, respectively. The CCL is less than 0.06 bps/Hz for the whole operating bandwidth.

Muhammad et al. [9] present a compact rectifier, capable of harvesting ambient radio frequency (RF) power. The total size of the rectifier is 45.4 mm × 7.8 mm × 1.6 mm, designed on FR-4 substrate using a single-stage voltage multiplier at 900 MHz. GSM/900 is among the favorable RF energy-harvesting (RFEH) energy sources that span over a wide range with minimal path loss and high input power. The proposed RFEH rectifier achieves measured and simulated RF-to-dc (RF to direct current) power conversion efficiency (PCE) of 43.6% and 44.3% for 0 dBm input power, respectively. Additionally, the rectifier attained 3.1-V DC output voltage across a 2-kΩ load terminal for 14 dBm and is capable of sensing low input power at −20 dBm.

Kamal et al. [10] propose and analyses a novel single-layer multiple-input multiple-output (MIMO) antenna for fifth-generation (5G) 28-GHz frequency band applications. The proposed MIMO antenna operates in the Ka-band, which is the most desirable frequency band for 5G mm-wave communication. The dielectric material is a Rogers-5880 with a relative permittivity, thickness, and loss tangent of 2.2, 0.787 mm, and 0.0009, respectively, in the proposed antenna design. The proposed MIMO configuration antenna element consists of triplet circular-shaped rings surrounded by an infinity-shaped shell. The simulated gain achieved by the proposed design is 6.1 dBi, while the measured gain is 5.5 dBi. Furthermore, the measured and simulated antenna efficiency is 90% and 92%, respectively. One of the MIMO performance metrics—i.e., the envelope correlation coefficient (ECC)—is also analyzed and found to be less than 0.16 for the entire operating bandwidth.

Bouknia et al. [11] illustrate and examine a theoretical study for the investigation of the electromagnetic field distributions and the input impedance of a printed dipole antenna structure loaded on a uniaxial anisotropic medium. The presentation of the electromagnetic field distributions, for which some examples have been shown here, provides a better understanding of the constitutive parameter (εt, εz, μt, and μz) contributions. Furthermore, the electrical and magnetic uniaxial anisotropy offers more degrees of freedom and further flexibility to realize a good direct matching effect on the input impedance. This shows that the complex media present a great potential in the design of innovative microwave components. This constitutes a starting point for understanding the behavior of the electromagnetic field in anisotropic and bianisotropic media, and many more interesting results are expected.

Addepalli et al. [12] design and implement UWB multiple-input multiple-output (MIMO) antennas with four and eight elements having connected grounds. The proposed antenna has a modified substrate geometry and comprises a circular arc-shaped conductive element on the top with the modified ground-plane geometry. Polarization diversity and isolation are achieved by replicating the elements, orthogonally forming a plus-shape antenna structure. The modified ground plane consists of an inverted L-strip and semi-ellipse slot over the partial ground that helps the antenna in achieving effective wide bandwidth spanning from (117.91%) 2.84–11 GHz. Both 4/8-port antennas help to achieve a size of 0.61 λ × 0.61 λ mm2 (lowest frequency), where the 4-port antenna is printed on the FR4 substrate. The 4-port UWB MIMO antenna attains a wide-impedance bandwidth, an omni-directional pattern, isolation of >15 dB, ECC of < 0.015, and an average gain of > 4.5 dB, making the MIMO antenna suitable for portable UWB applications.

Chung et al. [13] present an antenna structure with two branches that can achieve dual-band and broadband bandwidth characteristics. Moreover, the antenna performances were analyzed, validated, and manufactured. Thus, this design is suitable for in-vehicle infotainment and autopilot equipment systems in autonomous vehicle communication systems, including 5G, B5G, 4G, V2X, ISM band of WLAN, Bluetooth, WiFi 6 band, WiMAX, and Sirius/XM radio application.

Chung et al. [14] propose a small-slot antenna system (50 mm × 9 mm × 2.7 mm) for 4 × 4 multiple-input multiple-output (MIMO) on smart glasses devices. The antenna is set on the plastic temple, and the inverted F antenna radiates through the slot in the ground plane of the sputtered copper layer outside the temple. The antenna substrate is made of polycarbonate (PC), and its thickness is 2.7 mm (εr = 2.85, tan δ = 0.0092). The proposed antenna has a peak gain of 4.3 dBi and an antenna efficiency of 85.69% at 5.14 GHz. In addition, it also can obtain a peak gain of 3.3 dBi and antenna efficiency of 82.78% at 6.8 GHz. The measurement results show that this antenna has good performance, allowing future smart eyewear devices to be applied to Wi-Fi 5G (5.18–5.85 GHz) and Wi-Fi 6e (5.925–7.125 GHz).

We would like to take this opportunity to appreciate and thank all authors for their outstanding contributions and the reviewers for their fruitful comments and feedback. Special appreciation should also be paid to the Editorial Board of MDPI's *Electronics* journal for the opportunity to guest edit this Special Issue, and to the *Electronics* Editorial Office staff for their hard and precise work in maintaining a rigorous peer-review schedule and timely publication.

Funding: This research received no external funding.

Conflicts of Interest: The authors declare no conflict of interest.

References

1. Federal Communications Commission (FCC). The FCC's 5G FAST Plan, Spectrum. Available online: https://www.fcc.gov/5G (accessed on 15 December 2021).
2. Rappaport, T.S.; Sun, S.; Mayzus, R.; Zhao, H.; Azar, Y.; Wang, K.; Wong, G.N.; Schulz, J.K.; Samimi, M.; Gutierrez, F. Millimeter Wave Mobile Communications for 5G Cellular: It Will Work! *IEEE Access* **2013**, *1*, 335–349. [CrossRef]

3. Kim, Y.; Lee, H.; Hwang, P.; Patro, R.K.; Lee, J.; Roh, W.; Cheun, K. Feasibility of mobile cellular communications at millimeter wave frequency. *IEEE J. Sel. Top. Signal Processing* **2016**, *10*, 589–599. [CrossRef]
4. Tu, Y.; Al-Yasir, Y.I.A.; Ojaroudi Parchin, N.; Abdulkhaleq, A.M.; Abd-Alhameed, R.A. A Survey on Reconfigurable Microstrip Filter–Antenna Integration: Recent Developments and Challenges. *Electronics* **2020**, *9*, 1249. [CrossRef]
5. Al-Yasir, Y.I.A.; Alkhafaji, M.K.; Alhamadani, H.; Ojaroudi Parchin, N.; Elfergani, I.; Saleh, A.L.; Rodriguez, J.; Abd-Alhameed, R.A. A New and Compact Wide-Band Microstrip Filter-Antenna Design for 2.4 GHz ISM Band and 4G Applications. *Electronics* **2020**, *9*, 1084. [CrossRef]
6. Iqbal, A.; Altaf, A.; Abdullah, M.; Alibakhshikenari, M.; Limiti, E.; Kim, S. Modified U-Shaped Resonator as Decoupling Structure in MIMO Antenna. *Electronics* **2020**, *9*, 1321. [CrossRef]
7. Dahri, M.H.; Jamaluddin, M.H.; Seman, F.C.; Abbasi, M.I.; Ashyap, A.Y.I.; Kamarudin, M.R.; Hayat, O. A Novel Asymmetric Patch Reflectarray Antenna with Ground Ring Slots for 5G Communication Systems. *Electronics* **2020**, *9*, 1450. [CrossRef]
8. Altaf, A.; Iqbal, A.; Smida, A.; Smida, J.; Althuwayb, A.A.; Hassan Kiani, S.; Alibakhshikenari, M.; Falcone, F.; Limiti, E. Isolation Improvement in UWB-MIMO Antenna System Using Slotted Stub. *Electronics* **2020**, *9*, 1582. [CrossRef]
9. Muhammad, S.; Jiat Tiang, J.; Kin Wong, S.; Iqbal, A.; Alibakhshikenari, M.; Limiti, E. Compact Rectifier Circuit Design for Harvesting GSM/900 Ambient Energy. *Electronics* **2020**, *9*, 1614. [CrossRef]
10. Kamal, M.M.; Yang, S.; Ren, X.-C.; Altaf, A.; Kiani, S.H.; Anjum, M.R.; Iqbal, A.; Asif, M.; Saeed, S.I. Infinity Shell Shaped MIMO Antenna Array for mm-Wave 5G Applications. *Electronics* **2021**, *10*, 165. [CrossRef]
11. Bouknia, M.L.; Zebiri, C.; Sayad, D.; Elfergani, I.; Rodriguez, J.; Alibakhshikenari, M.; Abd-Alhameed, R.A.; Falcone, F.; Limiti, E. Theoretical Study of the Input Impedance and Electromagnetic Field Distribution of a Dipole Antenna Printed on an Electrical/Magnetic Uniaxial Anisotropic Substrate. *Electronics* **2021**, *10*, 1050. [CrossRef]
12. Addepalli, T.; Desai, A.; Elfergani, I.; Anveshkumar, N.; Kulkarni, J.; Zebiri, C.; Rodriguez, J.; Abd-Alhameed, R. 8-Port Semi-Circular Arc MIMO Antenna with an Inverted L-Strip Loaded Connected Ground for UWB Applications. *Electronics* **2021**, *10*, 1476. [CrossRef]
13. Chung, M.-A.; Yang, C.-W. Miniaturized Broadband-Multiband Planar Monopole Antenna in Autonomous Vehicles Communication System Device. *Electronics* **2021**, *10*, 2715. [CrossRef]
14. Chung, M.-A.; Hsiao, C.-W.; Yang, C.-W.; Chuang, B.-R. 4 × 4 MIMO Antenna System for Smart Eyewear in Wi-Fi 5G and Wi-Fi 6e Wireless Communication Applications. *Electronics* **2021**, *10*, 2936. [CrossRef]

Article

A New and Compact Wide-Band Microstrip Filter-Antenna Design for 2.4 GHz ISM Band and 4G Applications

Yasir I. A. Al-Yasir [1], Mohammed K. Alkhafaji [2], Hana'a A. Alhamadani [2],
Naser Ojaroudi Parchin [1], Issa Elfergani [3,*], Ameer L. Saleh [4], Jonathan Rodriguez [3] and
Raed A. Abd-Alhameed [1,5]

[1] Biomedical and Electronics Engineering, Faculty of Engineering and Informatics, University of Bradford, Bradford BD7 1DP, UK; Y.I.A.Al-Yasir@bradford.ac.uk (Y.I.A.A.-Y); N.OjaroudiParchin@bradford.ac.uk (N.O.P.); R.A.A.Abd@bradford.ac.uk (R.A.A.-A.)

[2] Electronic Techniques Department, Basra Technical Institute, Southern Technical University, Basra 61001, Iraq; M.khudhair@stu.edu.iq (M.K.A.); Hnaa.abdulrazzaq@stu.edu.iq (H.A.A.)

[3] Instituto de Telecomunicações, Campus Universitário de Santiago, 3810-193 Aveiro, Portugal; Jonathan@av.it.pt

[4] Department of Electrical Engineering, University of Misan, Misan 62001, Iraq; Ameer-lateef@uomisan.edu.iq

[5] Department of Communication and Informatics Engineering, Basra University College of Science and Technology, Basra 61004, Iraq

* Correspondence: I.t.e.elfergani@av.it.pt; Tel.: +351-234-377900

Received: 15 June 2020; Accepted: 28 June 2020; Published: 2 July 2020

Abstract: A new and compact four-pole wide-band planar filter-antenna design is proposed in this article. The effect of the dielectric material type on the characteristics of the design is also investigated and presented. The filter-antenna structure is formed by a fourth-order planar band-pass filter (BPF) cascaded with a monopole microstrip antenna. The designed filter-antenna operates at a centre frequency of 2.4 GHz and has a relatively wide-band impedance bandwidth of about 1.22 GHz and a fractional bandwidth (FBW) of about 50%. The effects of three different types of substrate material, which are Rogers RT5880, Rogers RO3003, and FR-4, are investigated and presented using the same configuration. The filter-antenna design is simulated and optimised using computer simulation technology (CST) software and is fabricated and measured using a Rogers RT5880 substrate with a height (h) of 0.81 mm, a dielectric constant of 2.2, and a loss tangent of 0.0009. The structure is printed on a compact size of 0.32 $\lambda_0 \times$ 0.30 λ_0, where λ_0 is the free-space wavelength at the centre frequency. A good agreement is obtained between the simulation and measurement performance. The designed filter-antenna with the achieved performance can find different applications for 2.4 GHz ISM band and 4G wireless communications.

Keywords: band-pass; compact; filter-antenna; LTE; microstrip; 4G

---

1. Introduction

The ever-increasing demand for compact wireless communication transceivers continues to impact microwave (MW) and radio frequency (RF) applications [1–10]. Some of the essential elements in such devices are the planar antennas and filters [11–20], where these components significantly affect the whole performance of the wireless communication system. In recent years, microstrip filter-antenna designs have become some of the most desired RF structures because of their low profile, compact size, lightweightness, and ease of fabrication [21–34]. The microstrip filtering antenna is also beneficial because it can be directly printed onto dielectric substrate materials [21]. Filtering antenna design has many applications, mostly in modern wireless communication systems where the filtering and radiation pattern responses are utilised simultaneously [22].

It is known that the use of a substrate material in the design of RF/microwave circuits is common and has some essential challenges. One of the design basics is to choose the appropriate substrate material type as well as the thickness to fit with a suitable application. Finding the dielectric substrate for printed circuit board (PCB) materials is a trade-off process between high-performance designs and the cost of these materials at the RF and MW frequencies. This represents a significant challenge for the designer. By recognising the key parameters and features of interest for PCB materials at high frequencies, such as how different types of PCBs behave with different kinds of substrate materials at high frequencies (millimetre-wave frequencies), the selection will be carefully conceived when choosing printed circuit board materials for use in such applications [23].

Recently, many microstrip filter-antenna designs using different types of substrate materials have been proposed [24–37]. In [24], a co-design of a filter-antenna using a multilayered-substrate is introduced for future wireless applications. The design consists of three-pole open-loop ring transmission lines and a T-shaped microstrip antenna. The multilayer technology is applied to achieve a compact size structure. A Rogers RT5880 substrate with a relative dielectric constant of 2.1 and a thickness of 0.5 is used in this structure. The filter-antenna design operates on 2.6 GHz with a fractional bandwidth of around 2.8% and has achieved a measured gain of 2.1 dB. The main advantage of this structure is the compact size, but it lacks simplicity in the construction due to the use of a multilayer substrate configuration. The filter-antenna presented in [25] followed the same procedure and achieved similar performance with a circular polarisation characteristic. However, the proposed configuration applied another design method based on the substrate integrated waveguide technology.

In [27], a dipole microstrip filter-antenna is presented with quasi-elliptic realised gain performance using parasitic resonators. The parasitic elements are designed based on the stepped-impedance resonators and utilised to generate two transmission zeros in the in-band transmission as well as two radiation nulls in the out-of-band bandwidth. The design is fabricated using an F4B-2 substrate with a dielectric constant of 2.4 and a thickness of 1.1 mm. The design also has an air layer located between the radiator and the ground layers with a height of 9 mm. The designed filter-antenna works at 1.85 GHz and has achieved a fractional bandwidth of 4.2%. The presented design offers not only good radiation in the passband region but also efficiently attenuates the noise signals in the stop-band spectrum.

Moreover, a wide-band balun filter-antenna design with a high roll-off skirt factor is presented in [30]. The model is composed of a fourth-order quasi-Yagi radiator cascaded with a multilayer balun microstrip filter. The balun filter is formed by five stepped impedance resonators, which improves the rejection ratio of the passband. The designed filter-antenna operates at 2.5 GHz with a fractional bandwidth of 22.9% and two transmission zeros at both edges of the passband. The design has achieved 5.4 dB realised gain with a high roll-off rejection level. Although the design presented some advantages such as wide bandwidth and high suppression level, it also required a multilayer substrate technology.

In this paper, a new and compact filter-antenna design with a wide-band performance for 2.4 GHz ISM band and 4G wireless communications is presented. Unlike other microstrip filter-antenna designs proposed in the literature, the design proposed in this paper has some advantages such as compact size, simple structure, high gain, and wide bandwidth with good S_{11} characteristics. Also, this work presents and investigates three different types of dielectric substrate materials used for the same filter-antenna configuration and also checks the obtained performance and its suitability for the application that it is designed for.

2. Properties of the Dielectric Substrates

FR-4 is a low-cost PCB material, made from fiberglass textile implemented in epoxy resin. The "FR" in FR-4 refers to "fire-resistant". It has mostly substituted the (flammable) board material G-10 because of this feature. The FR-4 material usually works well when designing below 1GHz. However, as frequencies rise beyond 1 GHz, the passive circuit elements have to be taken into consideration.

The main considerations for circuit design in the 3–6 GHz range involve skin effect, surface roughness, proximity effect, and dielectric substrate [38]. The FR-4 dielectric constant ε_r has been reported in the range 4.3–4.8 and is slightly dependent on the frequency. The loss tangent tanδ of FR-4 is 0.018.

Composite RT/Duroid 5880 microfiber reinforced PTFE is designed for demanding stripline and microstrip line applications. RT/Duroid is a glass microfiber reinforced PTFE (Poly Etra Fluoro Ethylene) composite built by Rogers corporation. It shows excellent chemical resistance, involving solvent and reagents utilised in printing and coating; ease of cutting, fabrication, shearing, and machining; and environment-friendliness. RT/Duroid 5880 has a low loss tangent tanδ of about 0.004 and dielectric constant $\varepsilon_r = 2.2$ [39]. RO3003 laminates are ceramic-filled (Poly Tetra Fluoro Ethylene) PTFE composite circuit materials with mechanical characteristics, which are uniform regardless of the choice of dielectric constant. This case allows the designers to develop multilayer board designs, which use different dielectric constant materials for several layers, without facing accuracy problems [40]. RO3003 has a low loss tangent (tanδ = 0.0013) and a dielectric constant ($\varepsilon_r = 3.0$).

3. Microstrip Filter-Antenna Configuration

The layout of the proposed filter-antenna structure is shown in Figure 1. The design consists of a four-pole band-pass filter (BPF) and a monopole patch antenna. The integration of the BPF with the monopole patch antenna is realised by connecting the second port of the BPF with the antenna. A 50-Ω microstrip transmission line is used to feed both the BPF and monopole patch antenna, so there is no need for an additional matching circuit, and so the size of the configuration is reduced. The four-pole band-pass filter is mainly composed of four resonators, which are connected to the microstrip feed-line established on a dielectric substrate material. The ground plane of the filter-antenna design has an L-shaped slot etched, as shown in Figure 2. Also, each resonator consists of a square open-loop ring with a longitudinal stripline ending with E-shaped arms. Therefore, a compact structure has been achieved with this configuration.

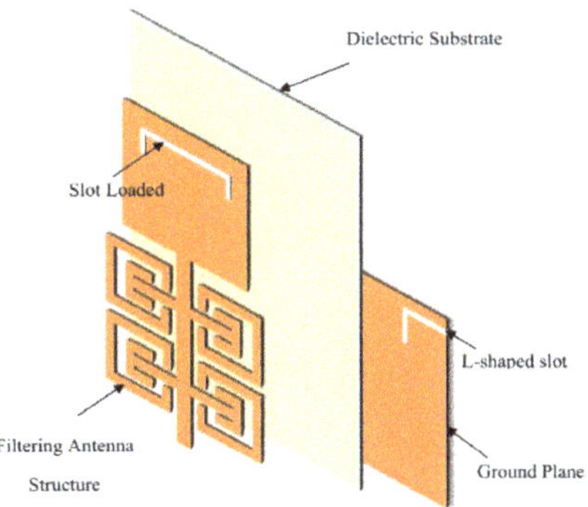

Figure 1. Filter-antenna structure layout.

The filtering antenna was established on three different types of dielectric substrate material, which are FR-4, RT/Duroid 5880, and RO3003. Firstly, these substrate materials were kept at a fixed thickness, and then the thickness of each dielectric substrate was changed to investigate which of these three types is more suitable for the specifications of a certain application. The absorption of electrical energy by a dielectric material that is exposed to an alternating electric field is called dielectric loss.

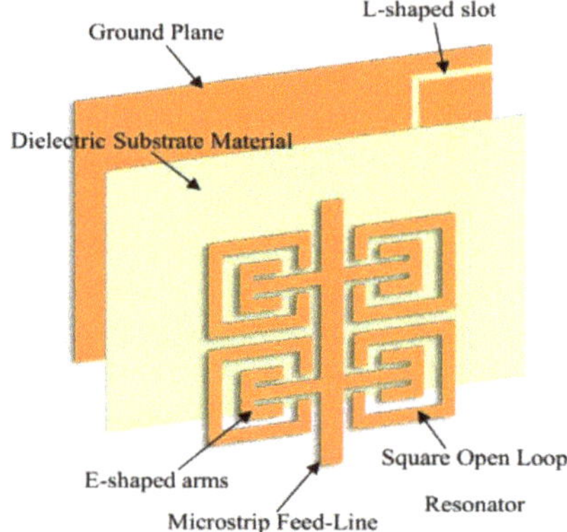

Figure 2. Four-pole band-pass filter structure layout.

Dielectric loss results from the influence of the limited loss tangent (tanδ), in which the losses are increasing and directly proportional to the operating frequency. Generally, the dielectric constant of the substrate ε_r is a complex number and is given by:

$$\varepsilon_r = \varepsilon_r' + j\varepsilon_r'' \tag{1}$$

where ε_r' is the real part of the dielectric constant and ε_r'' is the imaginary part of the dielectric constant. Then, the loss tangent is given by [41]:

$$\tan \delta = \frac{\varepsilon_r''}{\varepsilon_r'} \tag{2}$$

The relationship between the tangent loss and dielectric loss is given by the following formula [42]:

$$\alpha_d = \frac{|\varepsilon_r|}{\sqrt{\varepsilon_{reff}}} \cdot \frac{\varepsilon_{reff} - 1}{|\varepsilon_r| - 1} \cdot \frac{\pi}{\lambda_o} \cdot \tan \delta \tag{3}$$

where α_d is a dielectric loss, λ_o is a free-space wavelength, and ε_{reff} represents the effective dielectric constant of the substrate material and is given by:

$$\varepsilon_{reff} = \frac{\varepsilon_r + 1}{2} + \frac{\varepsilon_r - 1}{2}\left[1 + 12\,\frac{h}{W}\right]^{-1/2} \tag{4}$$

where W is the width of the patch.

Figure 3 shows the dielectric loss for the three types of substrates that are used in this article. The microstrip propagation delay t_{pd} is a function of a substrate dielectric constant ε_r and can be given by:

$$t_{pd}(\text{ns}/\text{cm}) = \left(\frac{1.017}{30.48}\right)\sqrt{0.475|\varepsilon_r| + 0.67} \tag{5}$$

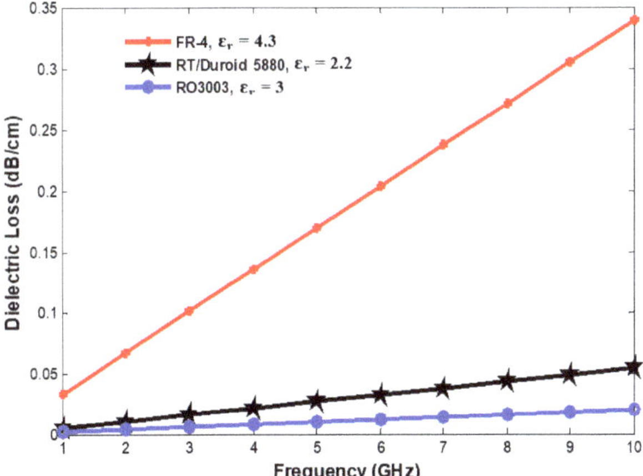

Figure 3. Dielectric loss versus frequency for different types of substrate materials at h/w = 0.04.

Attenuation sources of practical microstrip lines can be raised due to the effect of the radiation mechanism and the finite conductivity of the transmission lines. The energy in the microstrip line depends on the dielectric constant ε_r, substrate thickness (h), and the circuit geometry. Using a low dielectric constant substrate, which has low concentration energy, leads to high radiation losses. Figure 4 shows the relationship between the propagation delay and the dielectric constant for different types of dielectric substrates. Figure 5 shows the final filter-antenna structure with its optimised dimensions, as illustrated in Table 1. It can be seen that the structure has four open-loop resonators connected to the microstrip feed-line and established on a dielectric substrate material with a defected ground plane that has an L-shaped slot.

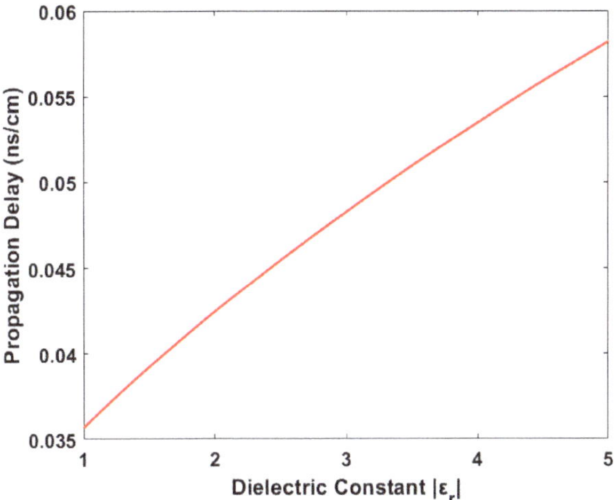

Figure 4. Propagation delay versus dielectric constant.

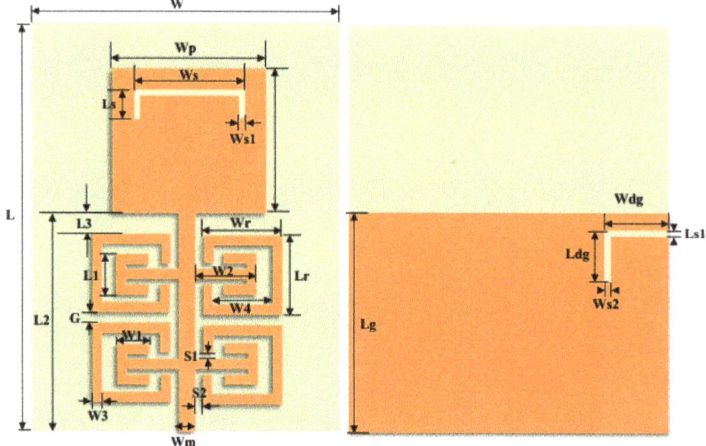

Figure 5. The proposed filter-antenna structure with its optimised dimensions.

Table 1. The optimised dimensions of the proposed filtering antenna (in mm).

Parameter	W	Wp	Ws	Ws1	Ws2	Wm	Wr	Wdg	W1	W2	W3	W4	W5
Dimensions	42	20	15	0.5	0.5	2.4	10	7.5	3.9	6.6	1	8	2.5
Parameter	L	Lg	Lr	Ls1	Ls1	L1	L2	L3	Lp	Ldg	S1	S2	G
Dimensions	45	25.4	9	0.4	3.5	4	26	3	18	5.4	0.5	0.8	1

4. Simulation and Measurement Results

The microstrip filter-antenna is designed at a centre frequency $f_0 = 2.4$ GHz, and bandwidth edges $f_1 = 1.80$ GHz and $f_2 = 3.10$ GHz. The proposed filter-antenna is suitable for 2.4 GHz ISM band and 4G wireless communications applications. Its relatively high bandwidth fits fast data transmission systems, which is required in modern and future wireless communications. Figure 6 shows the frequency response characteristics of the introduced filter-antenna without L-shaped defected ground structure (DGS).

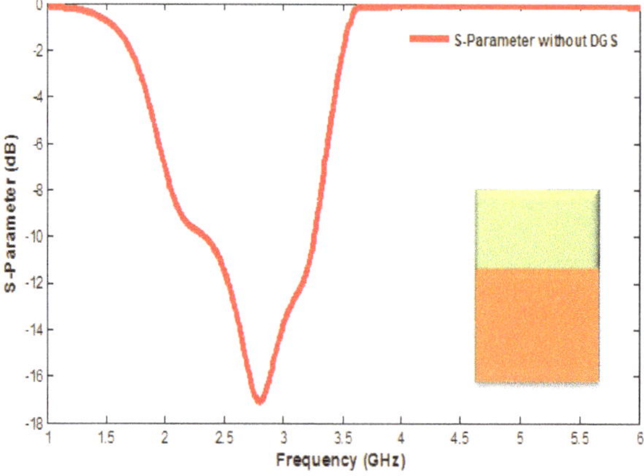

Figure 6. S_{11} of the proposed filter-antenna without defected ground structure (DGS).

Etching the DGS in the ground plane of the filter-antenna disturbs the current field distribution in a waveguide structure. This disturbance will affect the parameters of the design, such as the effective capacitance and effective inductance [43]. The feature of DGS is a slow-wave impact due to the equivalent LC components that may decrease the designed circuit size [44]. Figure 7 shows the simulated frequency response of the proposed filter-antenna design without a square open-loop ring resonator (SOLR).

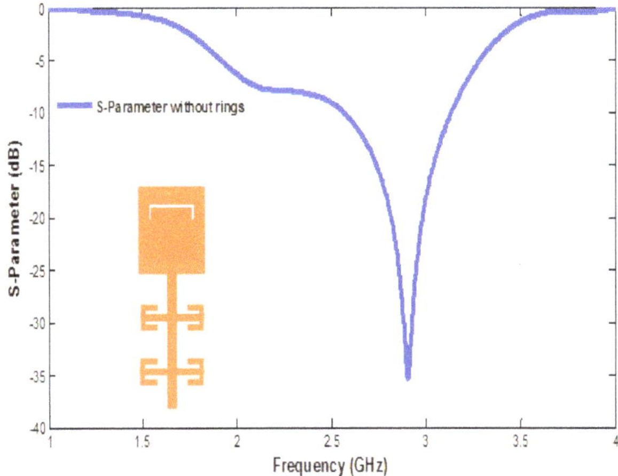

Figure 7. S_{11} of the filtering antenna without a square open-loop ring resonator (SOLR).

Furthermore, a performance comparison of the reflection coefficient parameters for three types of dielectric substrate materials at a fixed substrate height (h = 0.8 mm) is presented in Figure 8. It should be noted that the dielectric substrate material has a significant effect on the design performance, especially the centre frequency and reflection coefficient characteristics. Table 2 explains the performance comparison of the three different dielectric substrate types for which the filter-antenna is designed and proposed.

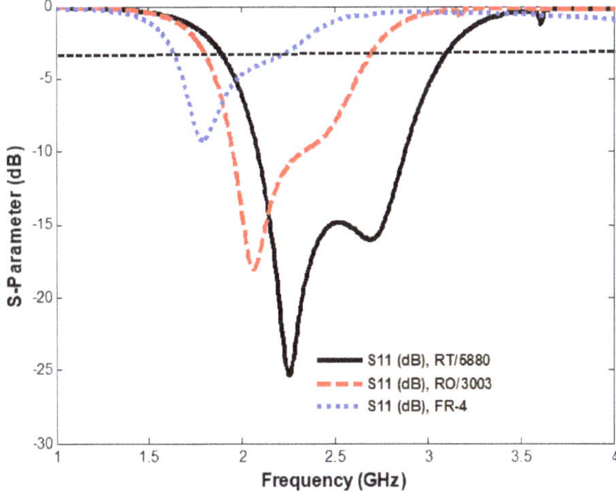

Figure 8. Comparison of S_{11} for different dielectric substrate materials at h = 0.81 mm.

Table 2. Comparison of different parameters for three different dielectric substrate materials.

Parameters	Dielectric Substrate Type		
	RT/5880	RO3003	FR-4
Centre frequency (f_0 GHz)	2.412	2.202	1.924
Return loss (dB)	15	12.065	6.0314
Maximum Gain (dB)	4.03	2.43	1.18
BW (GHz)	1.22	0.922	0.657
VSWR	1.1937	1.58	2.2

Figure 9 shows the comparison of the reflection coefficient parameters for three different dielectric substrate materials on which the filtering antenna design is established for different substrate heights. These comparisons are necessary to illustrate the effect of the dielectric substrate material type and thickness. Table 3 shows the comparison of some of the important parameters involved in the designed filter-antenna circuit on the three different dielectric substrate material types.

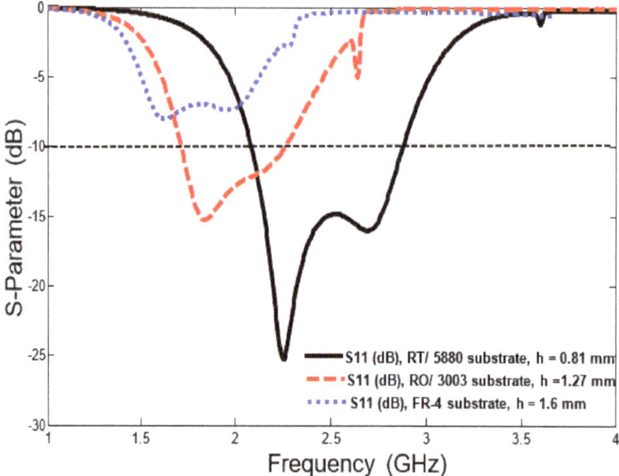

Figure 9. S_{11} comparison for different dielectric substrate materials and different dielectric substrate heights.

Table 3. Performance comparison of some of the parameters involved in the designed filter-antenna circuit using the three different dielectric substrate material types.

Parameters	Dielectric Substrate Properties		
	RT/5880 (h = 0.81 mm)	RO3003 (h = 1.27 mm)	FR-4 (h = 0.81 mm)
Centre frequency (GHz)	2.412	2.3	2.049
Return loss (dB)	15	13.063	7.04
Maximum Gain (dB)	4.1	2.63	2.22
BW (GHz)	1.22	1.3	1.059
VSWR	1.1937	1.52	2.24

Figure 10 shows the simulated and measured reflection coefficient (S_{11}) and gain for the proposed filter-antenna with the practical realisation for the prototype, which is fabricated using RT/Duroid 5880 dielectric substrate with a height of 0.81 mm and a dielectric constant of 2.2. The structure is printed on a compact size of $0.32\,\lambda_0 \times 0.30\,\lambda_0$, where λ_0 is the free-space wavelength at the centre frequency. From Figures 6, 7 and 10, it should be noted that there is a significant and noticeable effect of the DGS and the SOLR on the overall performance of the filter-antenna response. The designed filter-antenna operates at a centre frequency of 2.4 GHz and has a relatively wide-band impedance bandwidth of about 1.22 GHz and a fractional bandwidth (FBW) of about 50%. The measurement results show the design also has a maximum realised gain of 4.9 dB at the operating frequency.

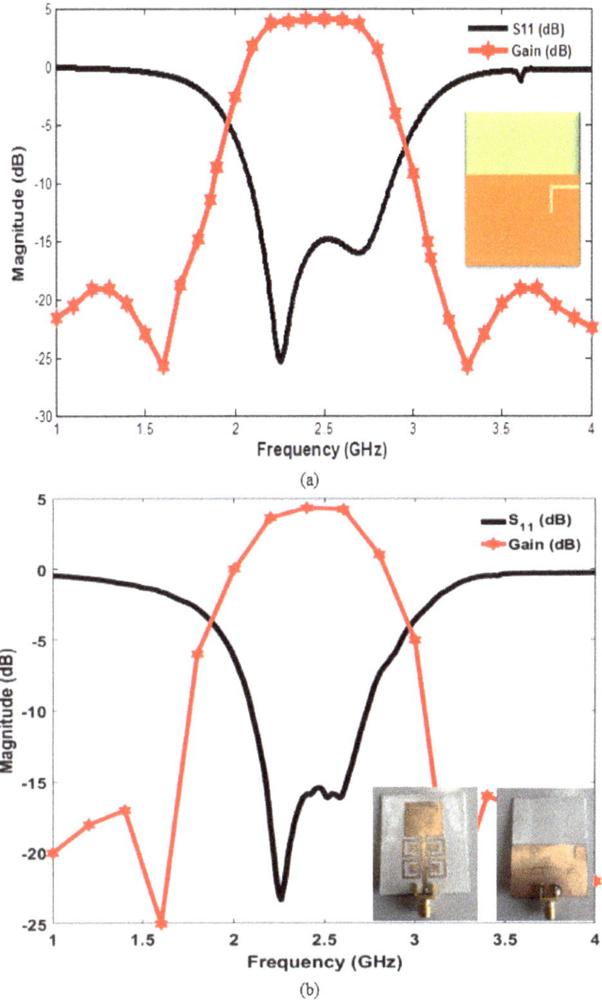

Figure 10. The performance of the proposed filter-antenna design: (**a**) simulated performance, (**b**) measured performance with a photograph of the fabricated prototype structure.

Moreover, Figure 11 shows the simulated and measured far-field radiation patterns of the proposed filter-antenna design at a centre frequency f_0 = 2.4 GHz. It is clear that the design offers stable and good radiation patterns at phi = 0 and 90 degrees. The simulation results from the CST simulator and the measurement results from the vector network analyzer (HP 8510C) and the anechoic chamber [41] show a reasonably good agreement.

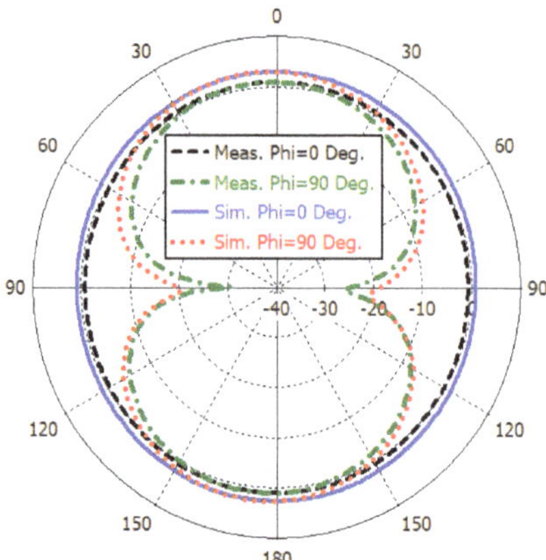

Figure 11. The far-field radiation patterns for the proposed filter-antenna design.

Table 4 shows a performance comparison between the proposed microstrip filter-antenna and some designs from the literature that have similar structures and performances. The proposed filter-antenna design has a compact size with a simple structure; it offers higher gain, wider fractional bandwidth, and good reflection coefficient characteristics.

Table 4. Comparison between the proposed design and others.

Ref.	Centre Frequency (GHz)	Fractional Bandwidth (%)	Size ($\lambda_0 \times \lambda_0$)	RL (dB)	Gain (dB)	Extra Structure
[24]	2.6	2.6	0.31×0.27	>13	2.2	Multilayer
[25]	11.65	4	2×1.1	>14	5.6	SIW *
[27]	1.85	5.4	0.74×0.74	>12	6.2	Multilayer
[28]	3.6	15	0.92×0.86	>14	10	Metasurface
[30]	2.5	22.8	1.7×1.3	>20	5	balun
[31]	2.5	15	0.76×0.76	>15	2	Multilayer
[32]	5	2	0.37×0.32	>15	4	None
[33]	2.45	6.4	0.72×0.70	>15	6	None
[35]	2.5	16.3	0.3×0.25	>20	2.4	None
[36]	2.5	8	0.45×0.45	>14	4.5	None
Prop.	2.4	50	0.32×0.30	>16	4.9	None

* Substrate integrated waveguide.

5. Conclusions

This article presents a compact wide-band microstrip filter-antenna design for 2.4 GHz ISM band and 4G wireless communications. The filter-antenna has been designed, measured, and studied in three different dielectric substrate materials, which are Rogers RT5880, Rogers RO3003, and FR-4. The analysis was performed by using CST microwave studio software. A performance comparison for the designed filter-antenna with different dielectric substrate materials and heights has been presented and discussed using the same design configuration. The results obtained from each design indicate that the most suitable characteristics for a specific application can be achieved by using Rogers RT5880 dielectric substrate material. According to the results, a filtering antenna that consists of a four-pole band-pass filter, integrated with a monopole patch antenna, was designed, fabricated, and measured. The simulation results generated by using the CST software package and the measurement achieved by using a vector network analyzer and an anechoic chamber show a reasonably good agreement.

Author Contributions: Conceptualisation, Y.I.A.A.-Y., H.A.A. and M.K.A.; methodology, Y.I.A.A.-Y. and N.O.P.; investigation, Y.I.A.A.-Y., M.K.A., I.E., J.R., A.L.S. and R.A.A.-A.; resources, Y.I.A.A.-Y., N.O.P, M.K.A. and R.A.A.-A.; writing—original draft preparation, Y.I.A.A.-Y., H.A.A., M.K.A., A.L.S., I.E., R.A.A.-A. and J.R.; writing—review and editing, Y.I.A.A.-Y., H.A.A., N.O.P. and I.E.; visualisation, Y.I.A.A.-Y., M.K.A., I.E., R.A.A.-A. and J.R. All authors have read and agreed to the published version of the manuscript.

Funding: This project has received funding from the European Union's Horizon 2020 research and innovation programme under grant agreement H2020-MSCA-ITN-2016 SECRET-722424.

Acknowledgments: The authors wish to express their thanks for the support provided by the innovation programme under grant agreement H2020-MSCA-ITN-2016 SECRET-722424.

Conflicts of Interest: The authors declare no conflict of interest.

References

1. Hussaini, A. Green Flexible RF for 5G. In *Fundamentals of 5G Mobile Networks*, 1st ed.; Rodriguez, J., Ed.; John Wiley and Sons: Hoboken, NJ, USA, 2015.

2. Rodriguez, J.; Radwan, A.; Barbosa, C.; Fitzek, F.H.P.; Abd-Alhameed, R.A.; Noras, J.M.; Jones, S.M.R.; Politis, I.; Galiotos, P.; Schulte, G.; et al. SECRET—Secure Network Coding for Reduced Energy next Generation Mobile Small Cells: A European Training Network in Wireless Communications and Networking for 5G. In Proceedings of the 2017 Internet Technologies and Applications (ITA), Wrexham, UK, 12–15 September 2017; pp. 329–333.

3. Parchin, N.O. *Microwave/RF Components for 5G Front-End Systems*; AVID SCIENCE: Telangana, India, 2020.

4. Al-Yasir, Y.; Abd-Alhameed, R.A.; Noras, J.M.; Abdulkhaleq, A.M.; Ojaroudi, N. Design of Very Compact Combline Band-Pass Filter for 5G Applications. In Proceedings of the Loughborough Antennas & Propagation Conference (LAPC), Loughborough, UK, 12–13 November 2018; pp. 1–4.

5. Abdulraheem, Y.I.; Abdullah, A.; Mohammed, H.; Abd-Alhameed, R.; Noras, J. Design of Frequency reconfigurable Multiband Compact Antenna using two PIN diodes for WLAN/WiMAX Applications. *IET Microw. Antennas Propag.* **2017**, *11*, 1098–1105. [CrossRef]

6. Al-Yasir, Y.I.A.; Tu, Y.; Bakr, M.S.; Parchin, N.O.; Asharaa, A.S.; Mshwat, W.A.; Abd-Alhameed, R.A.; Noras, J.M. Design of multi-standard single/tri/quint-wideband asymmetric stepped-impedance resonator filters with adjustable TZs. *IET Microw. Antennas Propag.* **2019**, *13*, 1637–1645. [CrossRef]

7. Liu, H.; Ren, B.P.; Li, S.; Guan, X.H.; Wen, P.; Peng, X.X.Y. High-Temperature Superconducting Bandpass Filter Using Asymmetric Stepped-Impedance Resonators with Wide-Stopband Performance. *IEEE Trans. Appl. Supercond.* **2015**, *25*, 1–6.

8. Al-Yasir, Y.I.A.; Tu, Y.; Parchin, N.O.; Abdulkhaleq, A.; Kosha, J.; Ullah, A.; Abd-Alhameed, R.; Noras, J. New Multi-standard Dual-Wideband and Quad-Wideband Asymmetric Step Impedance Resonator Filters with Wide Stop Band Restriction. *Int. J. RF Microw. Comput. Aided Eng.* **2019**, *29*, e21802. [CrossRef]

9. Al-Yasir, Y.I.A.; Tu, Y.; Parchin, N.O.; Elfergani, I.; Abd-Alhameed, R.; Rodriguez, J.; Noras, J. Mixed-coupling multi-function quint-wideband asymmetric stepped impedance resonator filter. *Microw. Opt. Tech. Lett.* **2019**, *61*, 1181–1184. [CrossRef]

10. Mohammed, H.J.; Abdulsalam, F.; Abdulla, A.S.; Ali, R.S.; Abd-Alhameed, R.A.; Noras, J.M.; Abdulraheem, Y.I.; Ali, A.; Rodriguez, J.; Abdalla, A.M. Evaluation of genetic algorithms, particle swarm optimisation, and firefly algorithms in antenna design. In Proceedings of the 13th International Conference on Synthesis, Modeling, Analysis and Simulation Methods and Applications to Circuit Design (SMACD), Lisbon, Portugal, 27–30 June 2016; pp. 1–4.

11. Hou, Z.; Liu, C.; Zhang, B.; Song, R.; Wu, Z.; Zhang, J.; He, D. Dual-/Tri-Wideband Bandpass Filter with High Selectivity and Adjustable Passband for 5G Mid-Band Mobile Communications. *Electronics* **2020**, *9*, 205. [CrossRef]

12. Al-Yasir, Y.I.A.; Ojaroudi Parchin, N.; Abdulkhaleq, A.M.; Bakr, M.S.; Abd-Alhameed, R.A. A Survey of Differential-Fed Microstrip Bandpass Filters: Recent Techniques and Challenges. *Sensors* **2020**, *20*, 2356. [CrossRef]

13. Al-Yasir, Y.I.A.; Parchin, N.O.; Abdulkhaleq, A.; Abd-Alhameed, R.; Noras, J. Recent Progress in the Design of 4G/5G Reconfigurable Filters. *Electronics* **2019**, *8*, 114. [CrossRef]

14. Ghouz, H.H.M.; Sree, M.F.A.; Ibrahim, M.A. Novel Wideband Microstrip Monopole Antenna Designs for WiFi/LTE/WiMax Devices. *IEEE Access* **2020**, *8*, 9532–9539. [CrossRef]

15. Lu, J.; Zhang, H.C.; He, P.H.; Zhang, L.P.; Cui, T.J. Design of Miniaturized Antenna Using Corrugated Microstrip. *IEEE Trans. Antennas Propag.* **2020**, *68*, 1918–1924. [CrossRef]

16. Al-Yasir, Y.; Abdullah, A.; Ojaroudi Parchin, N.; Abd-Alhameed, R.; Noras, J. A New Polarization-Reconfigurable Antenna for 5G Applications. *Electronics* **2018**, *7*, 293. [CrossRef]

17. Yasir, I.A.A.; Hasanain, A.H.A.; Baha, A.S.; Parchin, N.O.; Ahmed, M.A.; Abdulkareem, S.A.; Raed, A.A. New Radiation Pattern-Reconfigurable 60-GHz Antenna for 5G Communications. IntechOpen 2019. Available online: https://www.intechopen.com/online-first/new-radiation-pattern-reconfigurable-60-ghz-antenna-for-5g-communications (accessed on 26 September 2019).

18. Ogurtsov, S.; Koziel, S. A Conformal Circularly Polarized Series-Fed Microstrip Antenna Array Design. *IEEE Trans. Antennas Propag.* **2020**, *68*, 873–881. [CrossRef]

19. Al-Yasir, Y.I.A.; Parchin, N.O.; Abdulkhaleq, A.; Hameed, K.; Al-Sadoon, M.; Abd-Alhameed, R. Design, Simulation and Implementation of Very Compact Dual-band Microstrip Bandpass Filter for 4G and 5G Applications. In Proceedings of the 16th International Conference on Synthesis, Modeling, Analysis and Simulation Methods and Applications to Circuit Design (SMACD), Lausanne, Switzerland, 15–18 July 2019; pp. 41–44.

20. Yang, D.; Zhai, H.; Guo, C.; Li, H. A Compact Single-Layer Wideband Microstrip Antenna with Filtering Performance. *IEEE Antennas Wirel. Propag. Lett.* **2020**, *19*, 801–805. [CrossRef]

21. Khan, A.; Nema, R. Analysis of Five Different Dielectric Substrates on Microstrip Patch Antenna. *Int. J. Comput. Appl.* **2012**, *55*, 40–47. [CrossRef]

22. Chuang, C.-T.; Chung, S.-J. A new compact filtering antenna using defected ground resonator. In Proceedings of the Microwave Conference Proceedings (APMC), 2010 Asia-Pacific, Yokohama, Japan, 7–10 December 2010; pp. 1003–1006.

23. Coonrod, J. Choosing Circuit Materials for Millimeter-Wave Applications. *High Freq. Electron.* **2013**, *3*, 22–30.

24. Cui, J.; Zhang, A.; Yan, S. Co-design of a filtering antenna based on multilayer structure. *Int. J. RF Microw. Comput. Aided Eng.* **2020**, *30*, e22096. [CrossRef]

25. Hua, C.; Liu, M.; Lu, Y. Planar integrated substrate integrated waveguide circularly polarized filtering antenna. *Int. J. RF Microw. Comput. Aided Eng.* **2019**, *29*, e21517. [CrossRef]

26. Al-Yasir, Y.I.A.; Parchin, N.O.; Alabdallah, A.; Abdulkhaleq, A.M.; Abd-Alhameed, R.A.; Noras, J.M. Noras, Design of Bandpass Tunable Filter for Green Flexible RF for 5G. In Proceedings of the 2019 IEEE 2nd 5G World Forum (5GWF), Dresden, Germany, 30 September–2 October 2019.

27. Niu, B.; Tan, J.-H. Dipole filtering antenna with quasi-elliptic peak gain response using parasitic elements. *Microw. Opt. Technol. Lett.* **2019**, *61*, 1612–1616. [CrossRef]

28. Park, J.; Jeong, M.-J.; Hussain, N.; Rhee, S.; Park, S.; Kim, N. A low-profile high-gain filtering antenna for fifth generation systems based on nonuniform metasurface. *Microw. Opt. Technol. Lett.* **2019**, *61*, 2513–2519. [CrossRef]

29. Al-Yasir, Y.I.A.; Parchin, N.O.; Abd-Alhameed, R.A. A Differential-Fed Dual-Polarized High-Gain Filtering Antenna Based on SIW Technology for 5G Applications. In Proceedings of the 14th European Conference on Antennas and Propagation (EuCAP), Copenhagen, Denmark, 15–20 March 2020; pp. 1–5.

30. Song, L.; Wu, B.; Xu, M.; Su, T.; Lin, L. Wideband balun filtering quasi-Yagi antenna with high selectivity. *Microw. Opt. Technol. Lett.* **2019**, *61*, 2336–2341. [CrossRef]

31. Wu, W.; Fan, R.; Wang, J.; Zhang, Q. A broadband low profile microstrip filter-antenna with an omni-directional pattern. In Proceedings of the 2014 3rd Asia-Pacific Conference on Antennas and Propagation, Harbin, China, 26–29 July 2014; pp. 580–582.

32. Lin, C.; Chung, S. A Compact Filtering Microstrip Antenna with Quasi-Elliptic Broadside Antenna Gain Response. *IEEE Antennas Wirel. Propag. Lett.* **2011**, *10*, 381–384.

33. Wu, W.; Ma, B.; Wang, J.; Wang, C. Design of a microstrip antenna with filtering characteristics for wireless communication systems. In Proceedings of the 2017 Sixth Asia-Pacific Conference on Antennas and Propagation (APCAP), Xi'an, China, 16–19 October 2017; pp. 1–3.

34. Al-Yasir, Y.I.A.; Parchin, N.O.; Abd-Alhameed, R.A. New High-Gain Differential-Fed Dual-Polarized Filtering Microstrip Antenna for 5G Applications. In Proceedings of the 14th European Conference on Antennas and Propagation (EuCAP), Copenhagen, Denmark, 15–20 March 2020; pp. 1–5.

35. Wu, W.; Yin, Y.; Zuo, S.; Zhang, Z.; Xie, J. A New Compact Filter-Antenna for Modern Wireless Communication Systems. *IEEE Antennas Wirel. Propag. Lett.* **2011**, *10*, 1131–1134.

36. Ohira, M.; Ma, Z. An efficient design method of microstrip filtering antenna suitable for circuit synthesis theory of microwave band-pass filters. In Proceedings of the 2015 International Symposium on Antennas and Propagation (ISAP), Hobart, TAS, Australia, 9–12 November 2015; pp. 1–4.
37. Al-Yasir, Y.I.A.; Alhamadani, H.A.; Kadhim, A.S.; Ojaroudi Parchin, N.; Saleh, A.L.; Elfergani, I.T.E.; Rodriguez, J.; Abd-Alhameed, R.A. Design of a Wide-Band Microstrip Filtering Antenna with Modified Shaped Slots and SIR Structure. *Inventions* **2020**, *5*, 11. [CrossRef]
38. Leys, D. Best materials for 3–6 GHz design. *Print. Circuit Des. Manuf.* **2004**, 34–39.
39. Khare, A.; Nema, R.; Gour, P.; Thakur, R.K. New multiband E-shape microstrip patch antenna on RT DUROID 5880 substrate and RO4003 substrate for pervasive wireless communication. *Int. J. Comput. Appl.* **2010**, *975*, 8887. [CrossRef]
40. Afrin, S.S.; Dev, P.R. *Performance Analysis of Microstrip Patch Antenna for Ultra Wide Band Application*; East West University: Dhaka, Bangladesh, 2015.
41. Szostak, F.K.; Słobodzian, P. Broadband Dielectric Measurement of PCB and Substrate Materials by Means of a Microstrip Line of Adjustable Width. *IEEE Microw. Wirel. Compon. Lett.* **2018**, *1*, 945–947. [CrossRef]
42. Kapusuz, K.; Lemey, S.; Rogier, H. Substrate-Independent Microwave Components in Substrate Integrated Waveguide Technology for High-Performance Smart Surfaces. *IEEE Trans. Microw. Theory Tech.* **2018**, *1*, 3036–3047. [CrossRef]
43. Breed, G. An introduction to defected ground structures in microstrip circuits. *High Frequency Electron.* **2008**, *7*, 50–54.
44. Mahmud, F.S.; Razalli, M.S.; Rahim, H.A.; Hoon, W.F.; Ilyas, M.Z. Parametric studies on effects of defected ground structure (DGS) for dual band bandstop microstrip filter. *EPJ Web Conf.* **2017**, *1*, 1–5. [CrossRef]

Review

A Survey on Reconfigurable Microstrip Filter–Antenna Integration: Recent Developments and Challenges

Yuxiang Tu [1], Yasir I. A. Al-Yasir [1,*], Naser Ojaroudi Parchin [1], Ahmed M. Abdulkhaleq [1,2] and Raed A. Abd-Alhameed [1,3]

[1] Biomedical and Electronics Engineering, Faculty of Engineering and Informatics, University of Bradford, Bradford BD7 1DP, UK; ytu95@yahoo.com (Y.T.); n.ojaroudiparchin@bradford.ac.uk (N.O.P.); A.Abd@sarastech.co.uk (A.M.A.); r.a.a.abd@bradford.ac.uk (R.A.A.-A.)

[2] SARAS Technology Limited, Leeds LS12 4NQ, UK

[3] Department of Communication and Informatics Engineering, Basra University College of Science and Technology, Basra 61004, Iraq

* Correspondence: y.i.a.al-yasir@bradford.ac.uk; Tel.: +44-4115-5929-4

Received: 15 June 2020; Accepted: 29 July 2020; Published: 4 August 2020

Abstract: Reconfigurable and tunable radio frequency (RF) and microwave (MW) components have become exciting topics for many researchers and design engineers in recent years. Reconfigurable microstrip filter–antenna combinations have been studied in the literature to handle multifunctional tasks for wireless communication systems. Using such devices can reduce the need for many RF components and minimize the cost of the whole wireless system, since the changes in the performance of these applications are achieved using electronic tuning techniques. However, with the rapid development of current fourth-generation (4G) and fifth-generation (5G) applications, compact and reconfigurable structures with a wide tuning range are in high demand. However, meeting these requirements comes with some challenges, namely the increased design complexity and system size. Accordingly, this paper aims to discuss these challenges and review the recent developments in the design techniques used for reconfigurable filters and antennas, as well as their integration. Various designs for different applications are studied and investigated in terms of their geometrical structures and operational performance. This paper begins with an introduction to microstrip filters, antennas, and filtering antennas (filtennas). Then, performance comparisons between the key and essential structures for these aspects are presented and discussed. Furthermore, a comparison between several RF reconfiguration techniques, current challenges, and future developments is presented and discussed in this review. Among several reconfigurable structures, the most efficient designs with the best attractive features are addressed and highlighted in this paper to improve the performance of RF and MW front end systems.

Keywords: reconfigurable; tunable; radio frequency; filter; antenna; filter–antenna; filtenna; fourth generation (4G); fifth-generation (5G)

---

1. Introduction

The increasing demand for compact, simple, and efficient transceivers continues to impact the development of microwave (MW) and radio frequency (RF) applications [1–5]. Some of the essential elements in such devices are the planar antennas and filters [6–9], which significantly affect the whole performance of the wireless communication systems. Generally, RF interference is a big issue in the current and future wireless systems, such as the green RF front ends and wideband applications [10,11]. Microstrip bandpass filters (BPFs) are commonly used in several applications, mainly in RF and MW wireless communications, due to their effective role in suppressing interference and noise signals [12–14]. Recently, the office of communications (Ofcom) has identified a low bandwidth at 700 MHz, mid bandwidth (3.4–3.8 GHz), and an upper millimeter-wave bandwidth (24.25–27.5 GHz) for possible use with fifth-generation (5G) systems [15]. However, microstrip BPFs are utilized to attenuate the harmonic signals in fourth-generation (4G) and 5G applications [16–20]. For microstrip BPFs, the number of poles and zeros, input and output external quality factors, coupling coefficients, and the configuration of the resonators are important parameters that define the filter performance [21]. Most microstrip filter miniaturization approaches aim to analyze, control, or optimize these parameters [22]. Additionally, several design techniques have been introduced in the literature, such as stepped-impedance resonator (SIR), combline, open-ring, coupled-line, and stub impedance filters [23–27].

On the other hand, reconfigurability can also be utilized using RF electronic components, such as varactors and PIN diodes, which allow for the current distribution on the patches to be modified and then for the reflection coefficient and radiation patter characteristics to be adapted. Micro-electro-mechanical switches (MEMS) can also be considered, however these involve additional costs and extra circuits. In recent years, several reconfigurable microstrip BPFs have been introduced [28–36]. However, with the rapid development of current 4G and 5G applications, compact, efficient, and reconfigurable planar filters with a wide tuning range will be urgently needed [37].

In addition to reconfigurable microstrip filters, frequency-reconfigurable microstrip antennas have been investigated and developed for many years to provide important features to enhance the innovation and development of RF systems [38–41]. Another important factor to be considered by antenna designers and researchers these days, especially when designing antennas for mobile devices, is the geometrical size and design complexity of the RF elements. Therefore, antenna miniaturization techniques are continuously under review and study by many researchers and engineers. However, there are always new developments and updates in the literature related to these aspects. Due to the high demand for very small structures, the construction of more compact components is required, while the gain and radiation pattern properties should be maintained at the same time and for the same configuration [38]. Compact frequency-reconfigurable microstrip antennas have been introduced for several applications, such as mobile communication devices. Furthermore, these antennas are also needed for other applications, such as global systems for mobile communication (GSM), digital communication systems (DCS), personal communication systems (PCS), universal mobile telecommunication systems (UMTS), Bluetooth, wireless local area networks (LAN), and long-term evolution (LTE) [42–52].

In recent years, the microstrip filter–antenna integration designs have become some of the most desired structures because of their low profile, compact size, light weight, and ease of fabrication [53–71]. Microstrip filtering antennas are also beneficial because they can be printed directly onto the dielectric substrate materials [53]. Filtering antenna designs have many applications, mostly in modern wireless communication systems, where filtering and efficient radiation pattern responses can be obtained simultaneously [55]. Furthermore, reconfigurable microstrip filtering antennas have attracted increasing interest nowadays as they can deliver more efficient and multiple functionalities [72–86]. These designs do not implement microstrip antennas and filters separately, rather the filter is loaded onto the radiating patch instead, resulting in more compact structures and improving the entire performance of the RF and MW systems.

Few review papers discuss the reconfigurable filtering antenna designs that have been presented in the literature [87–89]. In [87,88], the papers focus on passive filtering antenna configurations with ultra-wideband characteristics. These papers do not present an extensive up-to-date review of the recent technologies utilized to implement the RF components (filters, antennas, and filtennas). In [89], a review of various integrated reconfigurable filter and antenna combinations was presented in 2015. Many design techniques have been investigated in recent years, achieving structures with compact sizes and simple configurations, which need further study.

Unlike other review papers, up-to-date reconfigurable microstrip filters and antennas and their integration are investigated in this paper by focusing on the latest development and design challenges for these components. According to the literature review carried out in this paper, performance comparisons between the key and essential reconfigurable structures are also presented and discussed. We point out the most efficient designs with the most attractive features for researchers and engineers for reconfigurable microstrip filters, antennas, and filtering antennas (filtennas). Additionally, Figure 1 shows a graphical summary of the reviewed design techniques in this paper. This manuscript is organized as follows. Section 2 discusses the latest updates in the reconfigurable microstrip filter design. Section 3 presents and reviews some efficient frequency-reconfigurable microstrip antennas. Section 4 surveys filter–antenna integration, as well as reconfigurable filtering antennas. All these sections are followed by performance comparisons to summarize the main characteristics and advantages for each structure. Section 5 provides a comparison between several RF reconfiguration switches. Section 6 presents the main challenges and recommendation for future research work. Finally, Section 6 summarizes the conclusions of our review.

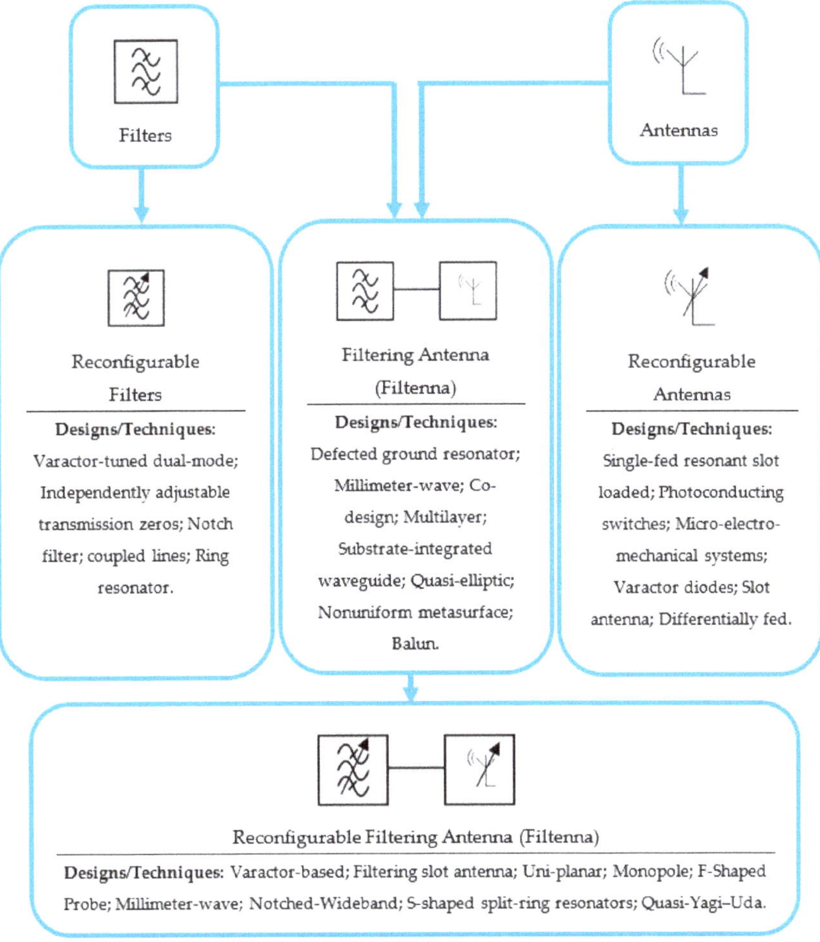

Figure 1. Graphical summary of the surveyed designs [28–86].

2. Reconfigurable Microstrip Filters

In recent years, several reconfigurable BPFs have been introduced [28–37]. A reconfigurable microstrip BPF using a varactor diode was designed and analyzed to achieve a constant impedance bandwidth in [29]. Reconfigurability is obtained by tuning the resonance frequencies for both the odd and even modes, where there is no mutual coupling between these two modes. Figure 2 shows the proposed tunable BPF with the obtained performance. The practical BPF performance depicts a good roll-off skirt on the low edge of the transmission band, with an insertion loss of less than 2.2 dB and a return loss of more than 10 dB. A 2.2–22.0 V reverse bias voltage is applied across the varactor diode to achieve a tuning rate of 40% for the 0.60–1.0 GHz range, with 91 MHz impedance bandwidth for all configurations.

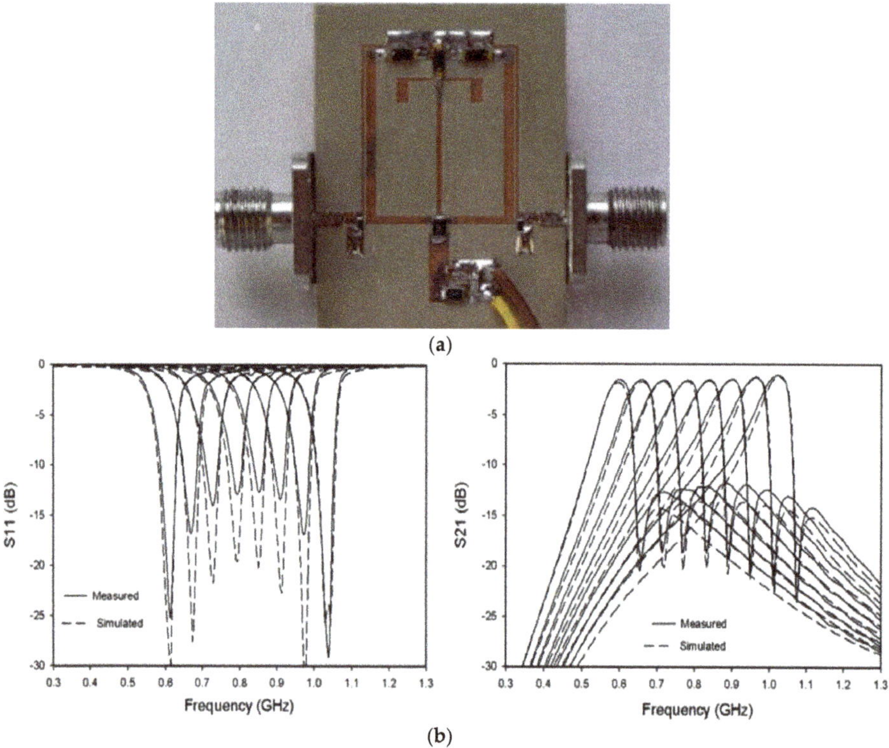

Figure 2. The reconfigurable filter reproduced from [29]. 2020, IEEE: (**a**) prototype structure; (**b**) S-parameter performance.

In [30], a reconfigurable microstrip BPF utilizes two varactors to tune two finite transmission zeros (TZs). The center frequency and the bandwidth are controlled to cover a wide range of about 600 MHz (1.4 GHz to 2.0 GHz) by altering the reverse bias voltage across the varactors (as seen in Figure 3). The measurement results show that the filter has an insertion loss of less than 4 dB, a return loss of more than 18 dB, and a fractional bandwidth of about 10%. A stopband rejection level of more than 25 dB is obtained by using the two transmission zeros. A 0.21–30.02 V bias voltage is applied across the diodes to tune the resonance frequency. In [31], a compact tunable planar BPF with a constant fractional bandwidth is introduced. By increasing the reverse bias voltage across the switches, the center frequency of the filter is tuned from 3.4 GHz to 3.8 GHz, with a fractional bandwidth of about 11%. The presented tunable filter has the advantages of having a compact size and simple structure, using only one varactor diode switch.

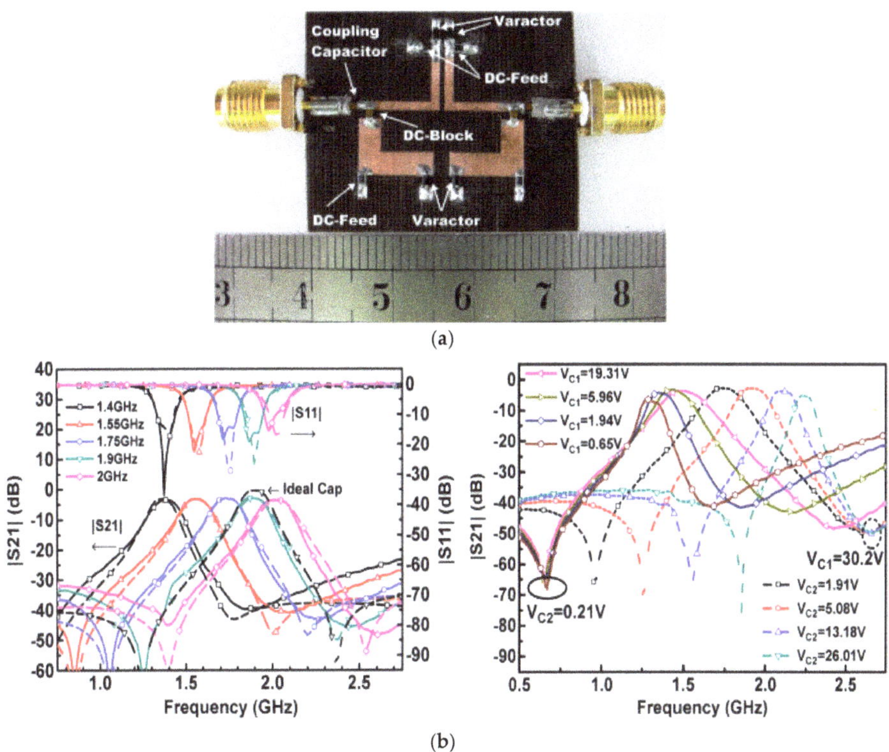

Figure 3. The reconfigurable filter reproduced from [30]. 2020, IEEE: (**a**) prototype structure; (**b**) S-parameter performance.

Ebrahimi et al. [32] proposed a notch dual-mode tunable bandstop planar filter using two varactor diodes. The proposed filter is implemented by loading inductive and capacitive coupling into the input and output transmission lines of the microstrip filter. The inductors were designed by using thin inductive strips. As illustrated in Figure 4, the second-order filter has a compact size of 0.13 λg × 0.17 λg and offers a continuous tuning range for the resonance frequency that ranges from 0.8 GHz to 1.1 GHz, with a stopband fractional bandwidth of about 17%. The measurement results show that the filter has 0.9 dB stopband return loss and 0.6 dB passband insertion loss over the entire tuning range. Apart from the other designs, the inductive coupling is achieved using an inductor in the bottom layer of the patch filter. This configuration avoids the need for a more complicated three-layered structure, provides more degrees of freedom in controlling the coupling coefficient factors, and maintains the top layer configuration, resulting in a more compact design.

Moreover, Chen et al. [33] introduced a 2-pole fully tunable planar filter with a small structure, continuous frequency tuning range, and constant impedance bandwidth. Two varactors are utilized to tune the resonance frequency between the high and low resonating modes. The tunable filter has a simple configuration that consists of a pair of reversed biased varactor diodes. Each resonator contains two transmission lines, which are connected together via a varactor diode. A 0.4–18 V bias voltage is applied to provide 0.3–2.4 pF capacitance. The tuning range for the resonance frequency was from 1.2 GHz to 1.9 GHz, with an operational impedance bandwidth of about 39 MHz. The proposed filter offers a compact size of 0.06 λg × 0.27 λg, continuous tunability, simple structure, and a wide-tuned spectrum, which make the designed BPF suitable for recent and future wireless communications. The proposed tunable filter with the achieved insertion and return losses is shown in Figure 5.

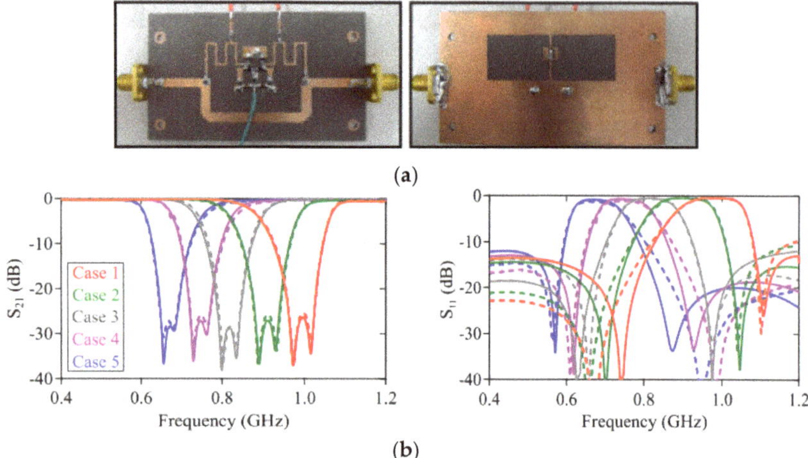

(a)

(b)

Figure 4. The reconfigurable filter reproduced from [32]. 2020, IEEE: (**a**) prototype structure; (**b**) S-parameter performance.

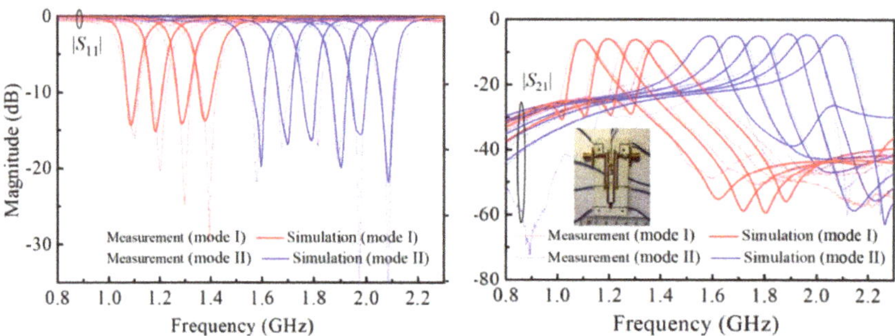

Figure 5. S-parameter performance of the reconfigurable filter reproduced from [33]. 2020, IEEE with a photograph of the fabricated prototype.

In [36], a very compact microstrip reconfigurable filter for fourth-generation (4G) and sub-6 GHz fifth-generation (5G) systems using a new hybrid co-simulation method is presented. The basic microstrip design uses three coupled line resonators with λ/4 open-circuit stubs. The coupling coefficients between the adjacent and non-adjacent resonators are used to tune the filter at the required center frequency to cover the frequency range of 2.5 GHz to 3.8 GHz. Figure 6 shows the simulated insertion and return losses of the proposed reconfigurable filter.

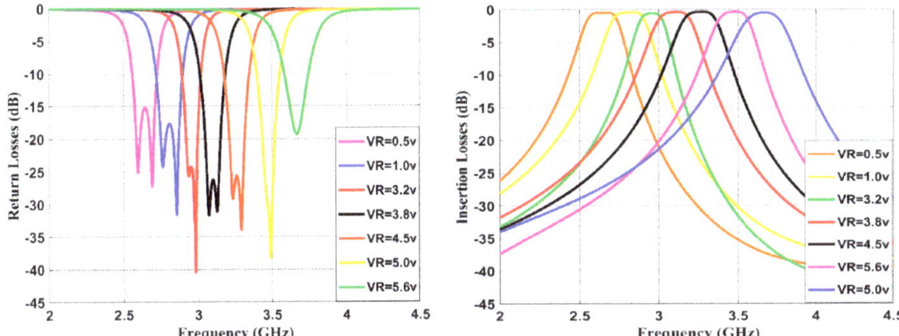

Figure 6. S-parameter performance of the reconfigurable filter reproduced from [36]. 2020, IEEE with a photograph of the prototype.

However, with the rapid development of current 4G and 5G applications, compact and reconfigurable planar filters with a wide tuning range are needed. To this end, several tunable filters have offered some attractive features that are essential for current and future wireless communications. Table 1 shows the comparative performance of the reviewed reconfigurable microstrip BPFs. It is clear that the proposed filter in [36] has a wider tuning range, wider impedance bandwidth, smaller insertion losses, and smaller size compared to the designs presented in [29,30,32–35]. The tunable filters presented in [29,33] have an impedance bandwidth of only 40 MHz. Additionally, the tunable filter proposed in [35,36] only use two varactor diode switches and a simple basing circuit to achieve the tunable frequency and efficient characteristics. As a result, the filter presented in [36] has very good performance in terms of the S-parameter group delay and the phase of S_{21}, along with other attractive features, such its compact size, relatively few tuning diodes, and simple structure; thus, it is a good option for many 5G systems.

Table 1. Performance comparison between the surveyed reconfigurable filters.

Ref.	Year	Topology	Tuning Range (GHz)	BW (MHz)	No. of Switches	IL * (dB)	Filter Size (mm³)	Challenges/ Limitations
[29]	2010	Dual-Mode	0.6–1.0	85–95	3	2.2	30 × 23 × 1.27	Low tuning range
[30]	2011	Coupled lines	1.5–2.0	110	4	4	36 × 30 × 0.80	High loss
[32]	2018	Dual-Mode	0.66–0.99	108	4	0.75	72 × 70 × 1.6	Low tuning range
[33]	2018	Ring-resonator	1.1–2.1	40	7	6	52 × 12 × 1.6	Number of switches
[34]	2018	Dual-Mode	1.7–2.9	40	7	4	36 × 35 × 0.8	Number of switches
[35]	2018	Multimode	0.76–2	75–150	2	1.2	100 × 8 × 0.50	Size
[36]	2019	Coupled lines	2.5–3.8	95–115	2	0.8	13 × 8 × 0.80	Constant bandwidth

* IL: Insertion loss.

3. Frequency-Reconfigurable Microstrip Antennas

This section focuses on the frequency-reconfigurable microstrip antennas. It introduces reconfigurable antennas with multislots distributed in the patch and ground in order to cover wireless local area network (WLAN) and worldwide interoperability for microwave access (WiMax) applications. Positive-intrinsic-negative (PIN) diode switches are used to change the effective electrical length of the antenna to cover the most important frequency ranges between 2 GHz to 6 GHz. Peroulis et al. demonstrated a tunable antenna using four PIN diode switches that change the effective length and S-shaped slot to operate in one of four selectable frequency bands ranges from 530 MHz to 890 MHz. Reconfiguration over such a wide frequency band is often accompanied by changes to the input impedance. However, the analyses of the antenna found the best position for the switches and adjusted the slot geometry such that the four frequency bands were obtained through the switching process, without a need to update the matching network or feed point position [42].

Panagamuwa et al. designed and proposed a balanced dipole antenna using a high-resistivity silicon. This design was equipped with two silicon photoconducting switches. Light from infrared laser diodes guided with fiber-optic cables was used to control the switches. When both switches are closed the antenna operates at a lower frequency of 2.16 GHz, while when both switches are open the antenna operates at 3.15 GHz. The researchers also noticed that the antenna gain changes with different optical power levels used to activate the switches [43], which is a disadvantage of this configuration.

Yang et al. proposed a U-slot frequency-reconfigurable microstrip antenna with a 50 Ω transmission line feed. By loading the slot to the radiating layer, flat and linear input impedance is achieved. Controlling the input impedance affects the operating frequency of the antenna. It has been shown that a trimmer can also adjust the input impedance of the microstrip antenna, such that the frequency ratio between the highest and lowest frequency is about 1.32 [45]. The presented reconfigurable antenna delivers a tuning range from 2.6 GHz to 3.35 GHz. On the other hand, Valkonen et al. presented a frequency-reconfigurable mobile terminal microstrip antenna using radio-frequency micro-electro-mechanical system (RF-MEMS) switches. The reconfigurability is obtained using a capacitive coupling element (CCE) to switch between two separate matching lines and then to adjust the state of the RF-MEMS switches [46]. The antenna is tunable between two configurations at 0.92 GHz and 1.8 GHz center frequencies. The design is printed on a PCB with a size of $24 \times 20 \times 3$ mm^3.

Moreover, Yu et al. introduced a very compact frequency-reconfigurable microstrip antenna with a very wide tuning range. Three varactor switches were used to provide tunable impedance. Using a new feeding technique, the obtained tunable frequency of the prototype design ranges from 458 MHz to 895 MHz, while the tuning bandwidth improvement was analyzed and discussed using the equivalent circuit parameters [47]. Figure 7 illustrates the prototype of the designed frequency tunable antenna with the achieved S-parameter performance.

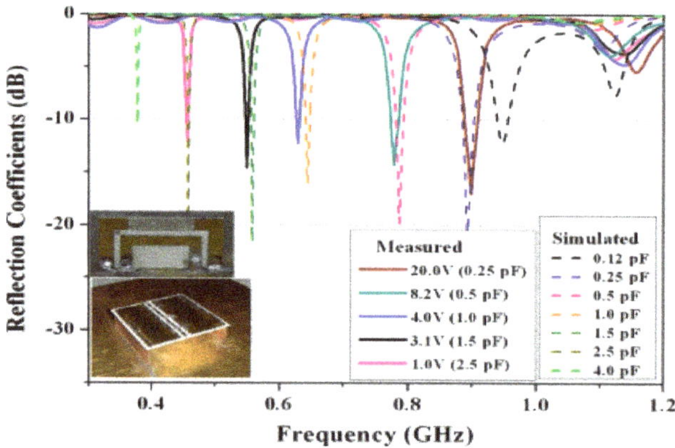

Figure 7. The reconfigurable antenna prototype and performance reproduced from [47]. 2020, IEEE.

Majid et al. introduced a compact, reconfigurable, frequency-agile, narrowband patch slot antenna. Six different center frequencies tunable from 2.1 GHz to 4.8 GHz were obtained in this design using five RF-PIN diode switches. To obtain the reconfigurability property, all the switches are placed in one slot, while the DC biasing circuit is built in the ground plane. The transmission line feeding circuit and the slot are bent to reduce about 35% of the original size of the structure, meaning a compact size is achieved [48]. In [49], Majid et al. also proposed a frequency-reconfigurable microstrip patch slot antenna using five RF-PIN diodes for cognitive wireless radio communications. Nine different operating frequencies covering the bandwidth from 2 GHz to 3.7 GHz are observed. To achieve the tunability property, the RF switches are also placed in the slot of the ground layer. Figure 8 shows a prototype of the designed antenna with the measured s-parameters.

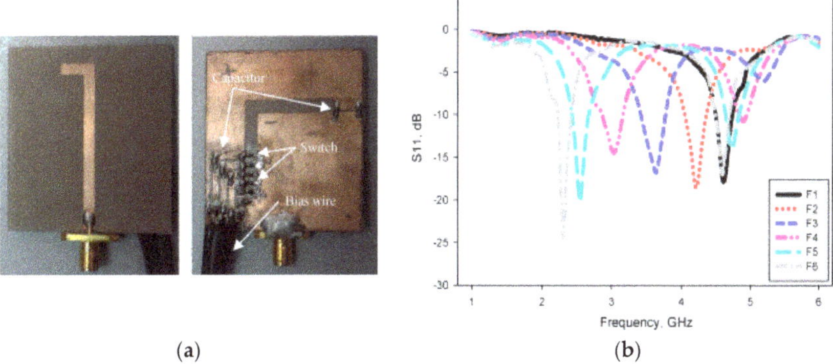

Figure 8. The reconfigurable antenna reproduced from [48]. 2020, IEEE: (**a**) prototype structure; (**b**) measured S-parameter results.

Recently, the new differential-fed technology was applied to design a frequency-reconfigurable microstrip antenna for sub-6 GHz 5G and WLAN wireless communications [50]. The antenna was designed based on pairs of vertical transmission lines to form two dipoles. Four RF-PIN diode switches are used to tune the antenna between 3.5 and 5.5 GHz. As seen in Figure 9, the proposed antenna offers impedance bandwidths of 2.9–4.2 GHz (fractional bandwidth of about 34%) and 5.0–6.2 GHz (fractional bandwidth of about 20%) for the two configurations for 5G and WLAN applications. The radiation pattern results are maintained for both configuration states. Table 2 compares the performance of this recently proposed technique with other studies from the literature. It should be noted that this technique offers excellent performance for the frequency-reconfigurable antenna designs, and thus is a good candidate for current and future wireless applications.

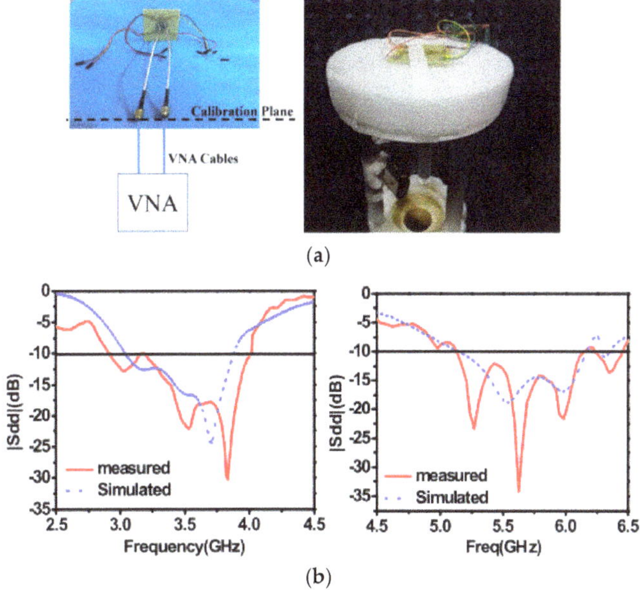

Figure 9. The reconfigurable antenna with the vector network analyzer (VNA) reproduced from [50]. 2020, IEEE: (**a**) prototype structure; (**b**) S-parameter performance for the two states.

Table 2. Performance comparison between the surveyed reconfigurable antennas.

Ref.	Year	Topology	Antenna Size (mm^3)	Tuning Range (GHz)	Type of Switches/DC Bias (V)	No. of Switches	No. of Achieved Bands	Constant Radiation Patterns (Challenges/ Limitations)
[45]	2008	U-Slot	$150 \times 150 \times 1.6$	2.6–3.35	Varactor (10.8–1.5)	1	6	No
[46]	2010	Inverted F	$40 \times 98 \times 5$	0.920–1.8	RF-MEMS (0.5–0.9)	1	2	No
[47]	2011	Capacitive loaded loop	$200 \times 200 \times 0.5$	0.45–0.89	Varactor (0.6–1.2)	3	5	No
[48]	2012	Patch slot	$50 \times 46 \times 1.6$	2.2–4.75	PIN Diode (0.9)	5	6	No
[49]	2013	Patch slot	$50 \times 50 \times 3.04$	1.98–3.59	PIN Diode (1.2)	5	9	No
[50]	2020	Differentially fed	$50 \times 50 \times 0.81$	2.9–6.2	PIN Diode (0.8)	4	2	Yes

It is shown that the designs presented in [45–49] provide variable radiation pattern characteristics for each state or band. This issue is one of the main challenges in the design of frequency-reconfigurable antennas, which has not been tackled yet for these structures. The structure presented in [50] not only offers a wide tuning range, but also keeps a constant radiation pattern performance over the tuned frequencies from 2.9 GHz to 6.2 GHz. The design presented in [49] has a smaller size than the antenna proposed in [50], despite this design using five PIN diodes. Nevertheless, the deigned antenna provides nine different bands with only five configurations, which makes the structure suitable for a wide range of wireless applications.

4. Microstrip Filter–Antenna (Filtenna) Integration

Recently many microstrip filter–antenna designs using different types of substrate materials have been proposed [53–71]. In [56], a co-design of a filter–antenna using a multilayered substrate is introduced for future wireless applications. The design consists of three-pole open-loop ring transmission lines and a T-shaped microstrip antenna. The multilayer technology is utilized to achieve a compact size structure. A Rogers RT5880 substrate with a relative dielectric constant of 2.1 and a thickness of 0.5 mm is used in this structure. The filter–antenna design operates at 2.6 GHz, with a fractional bandwidth of around 2.8% and a measured gain of 2.1 dB. While the main advantage of this structure is the compact size, it has a complex structure due to the use of a multilayer substrate configuration. The design presented in [57] also used the same design procedures and achieved similar performance, having a circular polarization characteristic. However, the filter–antenna design can involve different design techniques based on substrate-integrated waveguide (SIW) technology.

In [58], a dipole microstrip filter–antenna with quasi-elliptic gain performance using parasitic resonators is presented. The parasitic elements were designed based on the stepped-impedance resonators and utilized to generate two transmission zeros in the in-band transmission, as well as two radiation nulls in the out-of-band bandwidth. The design was fabricated using an F4B-2 substrate with a dielectric constant of 2.4 and a thickness of 1.1 mm. The design also has an air layer located between the radiator and the ground layers, with a height of 9 mm. The deigned filter–antenna works at 1.85 GHz and has a fractional bandwidth of 4.2%. The design offers not only good radiation in the passband region but it also efficiently attenuates the noise signals in the stopband spectrum. Moreover, a wideband balun filter–antenna design with a high roll-off skirt factor is presented in [61]. The design is composed of a fourth-order quasi-Yagi radiator cascaded with a multilayer balun microstrip filter. The balun filter is formed by five stepped impedance resonators, which improves the rejection ratio of the passband. The designed filter–antenna operates at 2.5 GHz with a fractional bandwidth of 22.9% and generates two transmission zeros at both edges of the passband. The design has achieved 5.4 dBi realized gain, with a high roll-off rejection level. Although the design has shown some advantages, such as the wide bandwidth and high suppression level, it also requires the use of multilayer substrate technology.

Recently, a very compact wideband microstrip filter antenna design with high gain and high selectivity was proposed in [71]. The design consists of a rectangular microstrip, four parasitic lines, two strip lines, and three shorting vias. The design is printed on an 80×80 mm^2 F-4B substrate with a dielectric constant of 2.6, loss tangent of 0.003, and a height of 4 mm. The center frequency of the design is 2.4 GHz, with an impedance bandwidth range of 2.19 GHz to 2.68 GHz (fractional bandwidth of 20.1%). The filter antenna has a realized gain of 9.5 dBi and flat radiation efficiency of more than 90%. Figure 10 shows the simulated and measured results with a prototype of the fabricated filtering antenna.

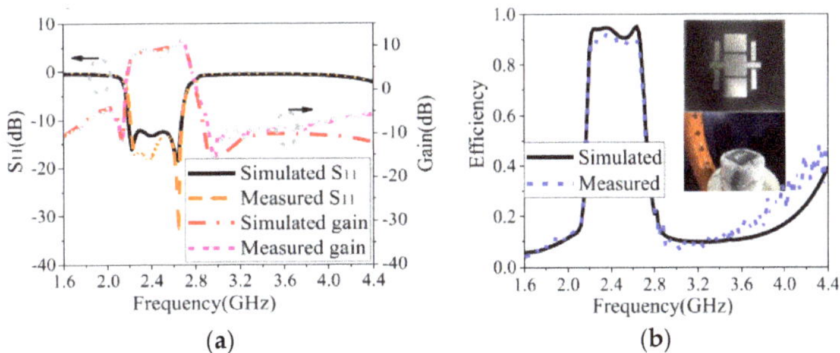

Figure 10. The filtering antenna design reproduced from [71]. 2020, IEEE: (**a**) S-parameter and gain; (**b**) efficiency and a photograph of the fabricated prototype.

However, design complexity and system size are other challenges facing designers of filtering antenna structures. As explained in the literature, many design approaches have been carried out to offer a simple structure and compact size, which can be easily integrated with other RF front end systems. The multilayer structures presented in [56,58,59,62] have not managed these requirements. Moreover, substrate integrated waveguide (SIW) technology and the balun configuration were other notable attempts, as presented in [57] and [61], respectively. To summarize these approaches, Table 3 shows the performance comparison between the surveyed microstrip filter–antenna designs from the literature, which have similar performance. It should be noted that the filter–antenna design proposed in [71] has a compact size with a simple structure and offers higher gain, higher selectivity, a wider fractional bandwidth, and good reflection coefficient characteristics. In summary, without a need for extra filtering circuits, the design presented in [71] offers a new solution for current and future filtering antenna designs.

Table 3. Comparison between the presented filter–antenna designs.

Ref.	Year	Topology	f_0 (GHz)	FBW (%)	Size ($\lambda_0 \times \lambda_0$)	RL (dB)	Gain (dBi)	Extra Structure (Challenges/Limitations)
[56]	2020	Coupled lines	2.6	2.6	0.31×0.27	> 13	2.2	Multilayer
[57]	2019	SIW	11.65	4	2×1.1	> 14	5.6	SIW
[58]	2019	Quasi-elliptic	1.85	5.4	0.74×0.74	> 12	6.2	Multilayer
[59]	2019	Patch slot	3.6	15	0.92×0.86	> 14	10	Metasurface
[61]	2016	Quasi-Yagi	2.5	22.8	1.7×1.3	> 20	5	balun
[62]	2014	Ring slot	2.5	15	0.76×0.76	> 15	2	Multilayer
[63]	2011	Quasi-elliptic	5	2	0.90×0.90	> 15	4	None
[64]	2017	Open-loop	2.45	6.4	0.72×0.70	> 15	6	None
[66]	2011	Coupled lines	2.5	16.3	0.70×0.70	> 20	2.4	None
[67]	2015	Ring slot	2.5	8	0.75×0.75	> 14	4.5	None
[71]	2020	Coupled lines	2.4	20.1	0.60×0.60	> 16	9.5	None

FBW: Fractional bandwidth; RL: Return loss; SIW: substrate integrated waveguide.

Additionally, many reconfigurable microstrip filter–antenna structures have been presented and discussed [72–86]. In [79], a multiband tunable filter cascaded with a monopole antenna for cognitive radio communications is presented. The reconfigurable design covers four useful applications, including 1.9 GHz (GSM), 2.5 GHz (Bluetooth), 3.6 GHz (WiMAX), and 5.3 GHz (WLAN). Additionally, the deigned multiband filter–antenna provides a gain range from 1.2 dBi to 3.5 dBi in the four operating bands, with small variations of about 0.5 dBi between the adjacent bands, delivering a radiation efficiency above 60%. Table 4 compares some of the similar reconfigurable filtering antenna designs in the literature with the design presented in [79]. However, it is shown that the reconfigurable filtering antenna presented in [79] has a smaller size and wide tuning range, covering four discrete configurations for four important wireless applications.

Table 4. Comparison between some reconfigurable filter–antenna designs.

Ref.	Year	Topology	Switches Number/Type	Size (mm)	Frequency Range (GHz)	Gain (dBi)	Advantages/Challenges/Limitations
[72]	2012	Hexagonal slot	1/Varactor	30 × 59	6.2–6.5	5.7–6.7	Band-limited control
[73]	2016	E-shaped patch	2/PIN diodes	36 × 14	2.1, 2.4	-	Dual-band only
[74]	2014	Slot resonator	2/PIN diodes	103 × 120	1.6–6	2.3	Large size
[75]	2017	Open-loop resonator	5/PIN diodes	40 × 45	2.2–11	2.1–2.3	Needs more diodes
[77]	2019	4 Distinct resonators	4/PIN diodes	30 × 60	1.8–5.2	1.1–3.4	Compact, discrete tuning

A filter–antenna design with a reconfigurable frequency and bandwidth using an F-shaped feeding network is presented in [77]. The new feeding technique generates a multipath coupling scheme and provides the cross-coupling required to improve the out-of-band characteristics. Additionally, two varactor diodes are used and designed within the feeding network. The achieved performance shows that the proposed reconfigurable filter–antenna design has tunable frequency ranges from 2 GHz to 2.52 GHz, a fractional bandwidth that is tunable from 2.2% to 21.3%, a measured maximum gain of about 7.6 dBi, and a measured peak total efficiency of 85%. Figure 11 shows a photograph of the implemented reconfigurable filtering antenna design with simulated and measured reflection coefficients and boresight gain. Table 5 presents the performance comparisons between some recently published reconfigurable filtering antenna designs.

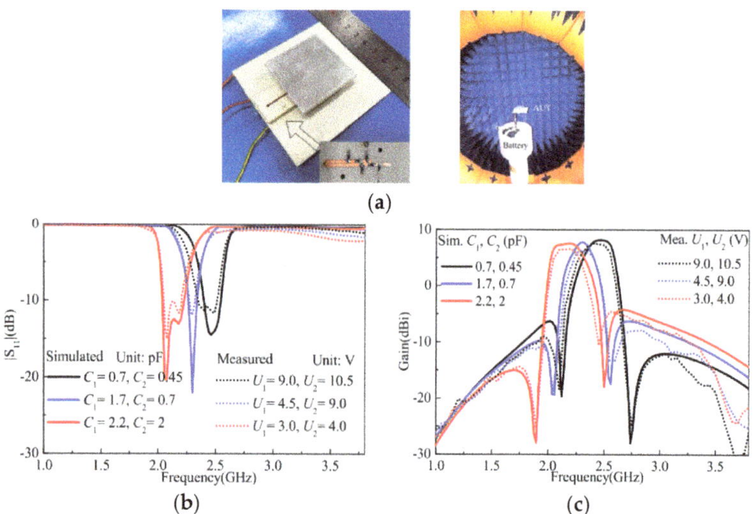

Figure 11. The reconfigurable filtering antenna reproduced from [77]. 2020, IEEE: (**a**) prototype structure; (**b**) S-parameter performance; (**c**) boresight gain performance.

Table 5. Performance comparisons between reconfigurable filter–antenna designs.

Ref.	Year	Topology	Size λ_0	Number of Switches	Frequency Range (GHz)	Gain (dBi)	Pattern Reconfiguration (Challenges/ Limitations)	Advantages
[81]	2015	Ring slot	$0.7 \times 0.3 \times 0.1$	1 PIN Diodes + 2 varactors	3.7–4.7	3	No	Wideband, tunable bandpass
[82]	2017	Coupled lines	$0.4 \times 0.2 \times 0.01$	2 PIN diodes	3–4.5	3.6	No	Wideband, tunable bandpass
[83]	2016	S-shaped split-ring	$0.4 \times 0.3 \times 0.002$	2 PIN diodes	3.1–3.8	1–2	No	Tunable bandpass, tunable bandstop
[84]	2019	Quasi-Yagi–Uda	$0.7 \times 0.7 \times 0.008$	2 PIN Diodes + 4 varactors	3.4–5.4	5–9	No	Tunable bandpass, tunable bandstop
[85]	2018	Coupled lines	$1.2 \times 1.2 \times 0.17$	4 PIN diodes	1.7–3.7	8–10	Yes	Wideband
[86]	2019	Coupled lines	$1.2 \times 1.6 \times 0.007$	4 PIN diodes	2.5–6.5	4.8	Yes	Wideband, tunable bandpass, tunable bandstop

It should be noted that considering both filter–antenna integration and reconfigurability properties at the same time will lead to some more advantages. However, this will also pose some challenges for both the biasing circuit and the structure configuration. In [48], two PIN diodes and four varactors are utilized in the basing circuit. Despite this configuration adding more complexity to the structure, it also results in a compact size and good performance in terms of the tuning range and the realized gain. It is also shown that wideband and tunable bandpass performance can be achieved by using the filter antenna integration design presented in [86]. This configuration has a high degree of freedom in terms of controlling the S-parameter characteristics and the radiation pattern behavior using a compact size structure. Thus, this makes the designed reconfigurable filter antenna a good candidate for current and future wireless applications.

5. Comparison between Switching Techniques

The common types of reconfiguration techniques that can be utilized to implement reconfigurable structures are illustrated in Table 6 [90–99]. Structures based on RF-MEMS [91], PIN diodes [92], and varactors [93] that redirect their surface currents are called "electrically reconfigurable." RF structures that use photoconductive configuration switch components are called "optically reconfigurable" [95]. Electronically reconfigurable or tunable elements are the best option when size and efficiency are required. However, the power handling capability and the lifetimes of these reconfiguration techniques cause some essential issues. PIN diodes operate in two configurations. The "on" state is where the diode is forward biased and the "off" state is where the diode is not biased or reverse-biased, while RF-MEMS uses mechanical movement to obtain a short circuit or an open circuit in the surface current path of RF elements. Unlike PIN diodes and RF MEMS, varactors can provide a continuous tuning range, with typical capacitance values range from tens to hundreds of picofarads. Moreover, unlike electrical reconfiguration, the photoconductive technique does not require the use of bias circuits and can be loaded in the RF PCB board without adding a complex design to modify the radiating elements. Additionally, the activation–deactivation mechanism for the switch does not create harmonic issues or intermodulation distortion. Conversely, in contrast with active switches, the optical switches are less common because of lossy characteristics and the need for complex activation approaches [99]. A description of the operation of the switches and comparisons between them are summarized in Table 6.

Table 6. Comparison between switching techniques [90–99].

Properties	PIN Diode	Varactor	RF MEMS	Photoconductive
Speed (μsec)	$1–100 \times 10^{-6}$	0.1	1–200	3–9
Quality factor	50–85	25–55	86–165	-
Voltage (V)	3–5	0.1–15	20–100	1.8–1.9
Current (mA)	3–20	1–25	0	0–87
Power (mW)	5–100	10–200	0.05-0.1	0–50
Temperature sensitivity	Medium	High	Low	Low
Cost	Low	Low	Medium	High
Loss at 1 GHz (dB)	0.3–1.2	0.5–3	0.05–0.2	0.5–1.5
Fabrication complexity	Commercially available	Commercially available	Low fabrication complexity	Complex

6. Current Challenges and Future Developments

Over the last few years, RF designers, researchers, and engineers have made a huge effort to explore reconfigurable filters and antennas and their integration as alternatives to the existing approaches and topologies, along with developing high-RF front end performance. Compared to the classical and passive filters and antennas, some essential challenges accompany the integrated and reconfigurable filters and antennas, which are efficient, compact, and multifunctional. Although recent researches show that microstrip planar configurations are capable of reducing the structure size, having the ability to produce a wider and flexible tuning range with low power and low loss is currently an important issue. As can be observed from the previous sections of this review, filter–antenna integration with reconfigurable characteristics requires a complex configuration, which can be considered as a common challenge for all reconfigurable transceivers. To overcome this challenge, some reconfigurable or tunable planar filters employing dual-mode ring resonators were introduced in [29,32,35]. Furthermore, the reconfigurable filter introduced in [36] has excellent performance in terms of the S-parameter group delay and S_{21} phase. Other features were also observed for this design, such as having a compact size, limited number of tuning diodes, and a simple structure.

Additionally, the realization of reconfiguration approaches in RF and MW components improves the multifunctional performance of the entire system. In the literature, several studies have stated the importance of reconfiguration techniques. For instance, an E-shaped microstrip wideband antenna with polarization diversity was presented in [100] to work in the frequency range of 2.3 GHz to 2.6 GHz. In a similar way, radiation pattern reconfigurable wideband microstrip antennas are also introduced in [101,102] to operate in the spectrum ranges of 2.3 GHz to 2.55 GHz and 1.6 GHz to 4 GHz, respectively. As shown in these papers, the integration of slots, lumped elements, and surface mount components in the radiating patch penetrates the radiation pattern performance. To overcome these problems, several papers in the literature utilize the feed line of the antenna to achieve filtering performance with reconfigurable characteristics. Some of the recent research studies in the literature that apply this technique to obtain filtering performance include [61–64]. Additionally, a filter–antenna design with a reconfigurable frequency and bandwidth using an F-shaped feeding network was presented in [103]. This technique generates a multipath coupling scheme and provides the cross-coupling required to improve the out-of-band characteristics.

Additionally, wideband filtering antenna designs are essential components of future wireless applications used to tackle high-speed and high data rate transmissions. For these designs, it is noticed that the size, insertion loss, and differential-mode bandwidth should also be taken into consideration and carefully investigated by the designers. Most of the introduced wideband and ultra-wideband filtering antenna configurations are designed based on a single-layer substrate. Therefore, it should be pointed out that using liquid crystal resonators and low-temperature co-fired ceramics can enhance the out-of-band rejection, thus improving and enhancing the performance of the wideband communication systems [104–106].

Reconfigurable filtering antennas based on substrate-integrated waveguide (SIW) technology can also be used for mmWave and 5G wireless communications to provide lower losses, higher quality factors, and more power handling capability when compared with the other surveyed approaches [107]. Additionally, using these techniques offers some advantages, such as enhancing the bandwidth and reducing the losses and sizes of the configurations. According to what is shown in this review, the design technique proposed in [86] can also overcome the challenges facing these technologies by using only one single-layer, half-mode, substrate-integrated waveguide resonator loaded with four slot lines. Furthermore, and with as any RF or microwave element, reconfigurable filters and antennas and systems combining both of these can also be designed, analyzed, and optimized using artificial intelligence, neural networks, and bio-inspired optimization algorithms [108–111]. These approaches can be utilized for future reconfigurable structures, since these designs require more analysis and parameter studies than classical and passive configuration. Therefore, using these approaches in the future could lead to overcoming several issues and challenges by processing many variables at one time. It is anticipated that new design techniques with high efficiency and fully reconfigurable characteristics will be seen shortly.

7. Conclusions

With the rapid development of 4G and 5G wireless communications in recent years, compact and reconfigurable or tunable structures with a wide tuning range have attracted more interest. Reconfigurable microstrip filters, antennas, and filter–antenna integration designs have been surveyed and discussed in this paper by focusing on the recent developments and challenges facing the researchers and engineers when dealing with these structures. It has been shown that integrating reconfigurable filters with the antennas can provide excellent interference suppression and maintain the fundamental radiation properties for the antennas. Performance comparisons between the main important reconfigurable designs have also been presented and discussed. The designs with the best performance were addressed and highlighted for possible future development and further studies to serve RF/MW front end systems. As seen in this paper, the reconfigurable filter proposed in [36] has a wider tuning range and a wider impedance bandwidth, smaller insertion losses, and a smaller size compared to the designs presented in [29,30,32–35]. As a reconfigurable antenna, the design presented in [49] has a smaller size than the antenna proposed in [50], despite this design using five PIN diodes. Nevertheless, the deigned antenna provides nine different bands with only five configurations, which makes the structure suitable for a wide range of wireless applications. It is also noted that wideband and tunable bandpass performance can be achieved by using the filter antenna integration design presented in [86]. The RF switches have also been discussed, summarized and compared. Finally, the paper has presented the current challenges and future developments for the three RF reconfigurable components, namely filters, antennas, and filtennas.

Author Contributions: Writing—original draft preparation, Y.T., Y.I.A.A.-Y., N.O.P., A.M.A., and R.A.A.-A.; writing—review and editing, Y.T., Y.I.A.A.-Y., and R.A.A.-A.; investigation, Y.T., Y.I.A.A.-Y., N.O.P., A.M.A., and R.A.A.-A.; resources, Y.I.A.A.-Y. and R.A.A.-A. For other cases, all authors have participated. All authors have read and agreed to the published version of the manuscript.

Funding: This project has received funding from the European Union's Horizon 2020 research and innovation program under grant agreement H2020-MSCA-ITN-2016 SECRET-722424.

Acknowledgments: The authors wish to express their thanks to the support provided by the innovation program under grant agreement H2020-MSCA-ITN-2016 SECRET-722424.

References

1. Hussaini, A.; Abdulraheem, Y.I.; Voudouris, K.N.; Mohammed, B.A.; Abd-Alhameed, R.A.; Mohammed, H.J.; Elfergani, I.; Abdullah, A.S.; Makris, D.; Rodriguez, J.; et al. Green Flexible RF for 5G. In *Fundamentals of 5G Mobile Networks*, 1st ed.; Rodriguez, J., Ed.; John Wiley and Sons: Hoboken, NJ, USA, 2015.

2. Al-Yasir, Y.I.A.; Tu, Y.; Bakr, M.S.; Parchin, N.O.; Asharaa, A.S.; Mshwat, W.A.; Abd-Alhameed, R.A.; Noras, J.M. Design of multi-standard single/tri/quint-wideband asymmetric stepped-impedance resonator filters with adjustable TZs. *IET Microw. Antennas Propag.* **2019**, *13*, 1637–1645. [CrossRef]

3. Liu, H.; Ren, B.P.; Li, S.; Guan, X.H.; Wen, P.; Peng, X.X.Y. High-Temperature Superconducting Bandpass Filter Using Asymmetric Stepped-Impedance Resonators with Wide-Stopband Performance. *IEEE Trans. Appl. Supercond.* **2015**, *25*, 1–6.

4. Al-Yasir, Y.I.A.; Tu, Y.; Parchin, N.O.; Abdulkhaleq, A.; Kosha, J.; Ullah, A.; Abd-Alhameed, R.; Noras, J. New Multi-Standard Dual-Wideband and Quad-Wideband Asymmetric Step Impedance Resonator Filters with Wide Stop Band Restriction. *Int. J. RF Microw. Comput. Aided Eng.* **2019**, *29*, 1–17. [CrossRef]

5. Tu, Y.; Guo, X.R.; Wang, C.H.; Jin, J. An improved 860–960 MHz fully integrated CMOS power amplifier designation for UHF RFID transmitter. *Int. J. Electron. Commun.* **2013**, *67*, 574–577. [CrossRef]

6. Ghouz, H.H.M.; Sree, M.F.A.; Ibrahim, M.A. Novel Wideband Microstrip Monopole Antenna Designs for WiFi/LTE/WiMax Devices. *IEEE Access* **2020**, *8*, 9532–9539. [CrossRef]

7. Lu, J.; Zhang, H.C.; He, P.H.; Zhang, L.P.; Cui, T.J. Design of Miniaturized Antenna Using Corrugated Microstrip. *IEEE Trans. Antennas Propag.* **2020**, *68*, 1918–1924. [CrossRef]

8. Ogurtsov, S.; Koziel, S. A Conformal Circularly Polarized Series-Fed Microstrip Antenna Array Design. *IEEE Trans. Antennas Propag.* **2020**, *68*, 873–881. [CrossRef]

9. Al-Yasir, Y.I.A.; Alkhafaji, M.K.; Alhamadani, H.A.; Ojaroudi Parchin, N.; Elfergani, I.; Saleh, A.L.; Rodriguez, J.; Abd-Alhameed, R.A. A New and Compact Wide-Band Microstrip Filter-Antenna Design for 2.4 GHz ISM Band and 4G Applications. *Electronics* **2020**, *9*, 1084. [CrossRef]

10. Hilt, A. Availability and Fade Margin Calculations for 5G Microwave and Millimeter-Wave Anyhaul Links. *Appl. Sci.* **2019**, *9*, 5240. [CrossRef]

11. Moghaddasi, J.; Wu, K. Multifunctional Transceiver for Future Radar Sensing and Radio Communicating Data-Fusion Platform. *IEEE Access* **2016**, *4*, 818–838. [CrossRef]

12. Hong, J.-S.; Lancaster, M.J. *Microstrip Filters for RF/Microwave Applications*; John Wiley and Sons: Hoboken, NJ, USA, 2004; Volume 167.

13. Richard, J.C.; Chandra, M.K.; Raafat, R.M. *Microwave Filters for Communication Systems Fundamentals, Design, and Applications*; John Wiley and Sons: Hoboken, NJ, USA, 2017.

14. Ian, H. *Theory and Design of Microwave Filters*; IET Electromagnetic Waves Series 48; IET: London, UK, 2006.

15. Statement: Improving Consumer Access to Mobile Services at 3.6 GHz to 3.8 GHz. Available online: https://www.ofcom.org.uk/consultations-and-statements/category-1/future-use-at-3.6-3.8-ghz (accessed on 21 October 2018).

16. Al-Yasir, Y.I.A.; Ojaroudi Parchin, N.; Abdulkhaleq, A.M.; Bakr, M.S.; Abd-Alhameed, R.A. A Survey of Differential-Fed Microstrip Bandpass Filters: Recent Techniques and Challenges. *Sensors* **2020**, *20*, 2356. [CrossRef] [PubMed]

17. Hou, Z.; Liu, C.; Zhang, B.; Song, R.; Wu, Z.; Zhang, J.; He, D. Dual-/Tri-Wideband Bandpass Filter with High Selectivity and Adjustable Passband for 5G Mid-Band Mobile Communications. *Electronics* **2020**, *9*, 205. [CrossRef]

18. Guan, Y.; Wu, Y.; Tentzeris, M.M. A Bidirectional Absorptive Common-Mode Filter Based on Interdigitated Microstrip Coupled Lines for 5G "Green" Communications. *IEEE Access* **2020**, *8*, 20759–20769. [CrossRef]

19. Al-Yasir, Y.I.A.; Parchin, N.O.; Abdulkhaleq, A.; Hameed, K.; Al-Sadoon, M.; Abd-Alhameed, R. Design, Simulation and Implementation of Very Compact Dual-band Microstrip Bandpass Filter for 4G and 5G Applications. In Proceedings of the 16th International Conference on Synthesis, Modeling, Analysis and Simulation Methods and Applications to Circuit Design (SMACD), Lausanne, Switzerland, 15–18 July 2019; pp. 41–44.

20. Al-Yasir, Y.I.A.; Parchin, N.O.; Alabdallah, A.; Abdulkhaleq, A.M.; Sajedin, M.; Elfergani, I.T.E.; Abd-Alhameed, R.A. Design, Simulation and Implementation of Very Compact Open-loop Trisection BPF for 5G Communications. In Proceedings of the 2019 IEEE 2nd 5G World Forum (5GWF), Dresden, Germany, 30 September–2 October 2019; pp. 189–193.

21. David, M.P. *Microwave Engineering*; John Wiley and Sons: Hoboken, NJ, USA, 2012.

22. Hotopan, R.; Cos, M.; Las-Heras, F. Reduced size C-band bandpass filter with 2nd harmonic suppression. In Proceedings of the 8th European Conference on Antennas and Propagation (EuCAP), The Hague, The Netherlands, 6–11 April 2014; pp. 971–974.

23. Ghatak, R.; Sarkar, P.; Mishra, R.K.; Poddar, D.R. A Compact UWB Bandpass Filter with Embedded SIR as Band Notch Structure. *IEEE Microw. Wirel. Compon. Lett.* **2011**, *21*, 261–263. [CrossRef]

24. Liu, H.; Liu, T.; Zhang, Q.; Ren, B.; Wen, P. Compact Balanced Bandpass Filter Design Using Asymmetric SIR Pairs and Spoof Surface Plasmon Polariton Feeding Structure. *IEEE Microw. Wirel. Compon. Lett.* **2018**, *28*, 987–989. [CrossRef]

25. Yuceer, M. A Reconfigurable Microwave Combline Filter. *IEEE Trans. Circuits Syst. II Express Briefs* **2016**, *63*, 84–88. [CrossRef]

26. Cho, Y.; Baek, H.; Lee, H.; Yun, S. A Dual-Band Combline Bandpass Filter Loaded by Lumped Series Resonators. *IEEE Microw. Wirel. Compon. Lett.* **2009**, *19*, 626–628. [CrossRef]

27. Velez, P.; Naqui, J.; Fernandez-Prieto, A.; Duran-Sindreu, M.; Bonache, J.; Martel, J.; Medina, F.; Martin, F. Differential Bandpass Filter With Common-Mode Suppression Based on Open Split Ring Resonators and Open Complementary Split Ring Resonators. *IEEE Microw. Wirel. Compon. Lett.* **2013**, *23*, 22–24. [CrossRef]

28. Al-Yasir, Y.I.A.; Parchin, N.O.; Abdulkhaleq, A.; Abd-Alhameed, R.; Noras, J. Recent Progress in the Design of 4G/5G Reconfigurable Filters. *Electronics* **2019**, *8*, 1–17. [CrossRef]

29. Tang, W.; Hong, J. Varactor-Tuned Dual-Mode Bandpass Filters. *IEEE Trans. Microw. Theory Tech.* **2010**, *58*, 2213–2219. [CrossRef]

30. Long, J.; Li Cui, C.; Huangfu, J.; Ran, L. A Tunable Microstrip Bandpass Filter with Two Independently Adjustable Transmission Zeros. *IEEE Microw. Wirel. Compon. Lett.* **2011**, *21*, 74–76. [CrossRef]

31. Al-Yasir, Y.I.A.; Parchin, N.O.; Alabdallah, A.; Abdulkhaleq, A.M.; Abd-Alhameed, R.A.; Noras, J.M. Design of Bandpass Tunable Filter for Green Flexible RF for 5G. In Proceedings of the 2019 IEEE 2nd 5G World Forum (5GWF), Dresden, Germany, 30 September–2 October 2019.

32. Ebrahimi, A.; Baum, T.; Scott, J.; Ghorbani, K. Continuously Tunable Dual-Mode Bandstop Filter. *IEEE Microw. Wirel. Compon. Lett.* **2018**, *28*, 419–421. [CrossRef]

33. Chen, C.; Wang, G.; Li, J. Microstrip Switchable and Fully Tunable Bandpass Filter with Continuous Frequency Tuning Range. *IEEE Microw. Wirel. Compon. Lett.* **2018**, *28*, 500–502. [CrossRef]

34. Chen, F.; Li, R.; Chen, J. Tunable Dual-Band Bandpass-to-Bandstop Filter Using p-i-n Diodes and Varactors. *IEEE Access* **2018**, *6*, 46058–46065. [CrossRef]

35. Lu, D.; Tang, X.; Barker, N.; Feng, Y. Single-Band and Switchable Dual-/Single-Band Tunable BPFs With Predefined Tuning Range, Bandwidth, and Selectivity. *IEEE Microw. Wirel. Compon. Lett.* **2018**, *66*, 1215–1227. [CrossRef]

36. Al-Yasir, Y.; Parchin, N.O.; Rachman, Z.-A.S.A.; Ullah, A.; Abd-Alhameed, R. Compact tunable microstrip filter with wide-stopband restriction and wide tuning range for 4G and 5G applications. In Proceedings of the IET's Antennas and Propagation Conference, Birmingham, UK, 11–12 November 2019; pp. 1–6.

37. Iqbal, A.; Tiang, J.J.; Lee, C.K.; Mallat, N.K.; Wong, S.W. Dual-Band Half Mode Substrate Integrated Waveguide Filter With Independently Tunable Bands. *IEEE Trans. Circuits Syst. II Express Briefs* **2020**, *67*, 285–289. [CrossRef]

38. Bernhard, J. *Reconfigurable Antennas*; Morgan and Claypool: San Rafael, CA, USA, 2007.

39. Abdulraheem, Y.I.; Abdullah, A.; Mohammed, H.; Abd-Alhameed, R.; Noras, J. Design of Frequency-reconfigurable Multiband Compact Antenna using two PIN diodes for WLAN/WiMAX Applications. *IET Microw. Antennas Propag.* **2017**, *11*, 1098–1105. [CrossRef]

40. Al-Yasir, Y.; Abdullah, A.; Ojaroudi Parchin, N.; Abd-Alhameed, R.; Noras, J. A New Polarization-Reconfigurable Antenna for 5G Applications. *Electronics* **2018**, *7*, 293. [CrossRef]

41. Al-Yasir, Y.I.A.; Abdullah, A.; Mohammed, H.; Mohammed, B.; Abd-Alhameed, R. Design of Radiation Pattern-Reconfigurable 60-GHz Antenna for 5G Applications. *J. Telecommun.* **2014**, *27*, 1–6.

42. Peroulis, D.; Sarabandi, K.; Katehi, L.P.B. Design of reconfigurable slot antennas. *IEEE Trans. Antennas Propag.* **2005**, *53*, 645–654. [CrossRef]

43. Panagamuwa, C.J.; Chauraya, A.; Vardaxoglou, J.C. Frequency and beam reconfigurable antenna using photoconducting switches. *IEEE Trans. Antennas Propag.* **2006**, *54*, 449–454. [CrossRef]

44. Yasir, I.A.A.; Hasanain, A.H.A.; Baha, A.S.; Parchin, N.O.; Ahmed, M.A.; Abdulkareem, S.A.; Raed, A.A. New Radiation Pattern-Reconfigurable 60-GHz Antenna for 5G Communications. Available online: https://www.intechopen.com/online-first/new-radiation-pattern-reconfigurable-60-ghz-antenna-for-5g-communications (accessed on 26 September 2019).

45. Yang, S.-L.S.; Kishkand, A.A.; Lee, K.-F. Frequency Reconfigurable U-Slot Microstrip Patch Antenna. *IEEE Antennas Wirel. Propag. Lett.* **2008**, *7*, 127–129. [CrossRef]

46. Valkonen, R.; Luxey, C.; Holopainen, J.; Icheln, C. Frequency-reconfigurable mobile terminal antenna with MEMS switches. In Proceedings of the Fourth European Conference on Antennas and Propagation, Barcelona, Spain, 12–16 April 2010; pp. 1–5.

47. Yu, Y.; Xiong, J.; Li, H.; He, S. An Electrically Small Frequency Reconfigurable Antenna with a Wide Tuning Range. *IEEE Antennas Wirel. Propag. Lett.* **2011**, *10*, 103–106.

48. Majid, H.A.; Rahim, M.K.A.; Hamid, M.R.; Ismail, M.F. A Compact Frequency-Reconfigurable Narrowband Microstrip Slot Antenna. *IEEE Antennas Wirel. Propag. Lett.* **2012**, *11*, 616–619. [CrossRef]

49. Majid, H.A.; Rahim, M.K.A.; Hamid, M.R.; Murad, N.A.; Ismail, M.F. Frequency-Reconfigurable Microstrip Patch-Slot Antenna. *IEEE Antennas Wirel. Propag. Lett.* **2013**, *12*, 218–220. [CrossRef]

50. Jin, G.; Deng, C.; Xu, Y.; Yang, J.; Liao, S. Differential Frequency-Reconfigurable Antenna Based on Dipoles for Sub-6 GHz 5G and WLAN Applications. *IEEE Antennas Wirel. Propag. Lett.* **2020**, *19*, 472–476. [CrossRef]

51. Al-Yasir, Y.I.A.; Parchin, N.O.; Elfergani, I.; Abd-Alhameed, R.A.; Noras, J.M.; Rodriguez, J.; Al-jzari, A.; Hammed, W.I. A New Polarization-Reconfigurable Antenna for 5G Wireless Communications. In *Broadband Communications, Networks, and Systems. BROADNETS 2018. Lecture Notes of the Institute for Computer Sciences, Social Informatics and Telecommunications Engineering*; Sucasas, V., Mantas, G., Althunibat, S., Eds.; Springer: Cham, Germany, 2019; Volume 263.

52. Al-Yasir, Y.I.A.A.; JaroudiParchin, N.O.; Alabdullah, A.; Mshwat, W.; Ullah, A.; Abd-Alhameed, R. New Pattern Reconfigurable Circular Disk Antenna Using Two PIN Diodes for WiMax/WiFi (IEEE 802.11 a) Applications. In Proceedings of the 2019 16th International Conference on Synthesis, Modeling, Analysis and Simulation Methods and Applications to Circuit Design (SMACD), Lausanne, Switzerland, 15–18 July 2019; pp. 53–56.

53. Khan, A.; Nema, R. Analysis of five different dielectric substrates on microstrip patch antenna. *Int. J. Comput. Appl.* **2012**, *55*, 44–47. [CrossRef]

54. Chuang, C.-T.; Chung, S.-J. A new compact filtering antenna using defected ground resonator. In *2010 Asia-Pacific Microwave Conference (APMC)*; IEEE: Piscataway Township, NJ, USA, 2010; pp. 1003–1006.

55. Coonrod, J. Choosing Circuit Materials for Millimeter-Wave Applications. *High Freq. Electron.* **2013**, *1*, 22–30.

56. Cui, J.; Zhang, A.; Yan, S. Co-design of a filtering antenna based on multilayer structure. *Int. J. RF Microw. Comput. Aided Eng.* **2020**, *30*, e22096. [CrossRef]

57. Hua, C.; Liu, M.; Lu, Y. Planar integrated substrate integrated waveguide circularly polarized filtering antenna. *Int. J. RF Microw. Comput. Aided Eng* **2019**, *29*, e21517. [CrossRef]

58. Niu, B.-J.; Tan, J.-H. Dipole filtering antenna with quasi-elliptic peak gain response using parasitic elements. *Microw. Opt. Technol. Lett.* **2019**, *61*, 1612–1616. [CrossRef]

59. Park, J.; Jeong, M.; Hussain, N.; Rhee, S.; Park, S.; Kim, N. A low-profile high-gain filtering antenna for fifth generation systems based on nonuniform metasurface. *Microw. Opt. Technol. Lett.* **2019**, *61*, 2513–2519. [CrossRef]

60. Al-Yasir, Y.I.A.; Parchin, N.O.; Abd-Alhameed, R.A. A Differential-Fed Dual-Polarized High-Gain Filtering Antenna Based on SIW Technology for 5G Applications. In Proceedings of the 14th European Conference on Antennas and Propagation (EuCAP), Copenhagen, Denmark, 15–20 March 2020; pp. 1–5.

61. Song, L.; Wu, B.; Xu, M.; Su, T.; Lin, L. Wideband balun filtering quasi-Yagi antenna with high selectivity. *Microw. Opt. Technol. Lett.* **2019**, *61*, 2336–2341. [CrossRef]

62. Wu, W.; Fan, R.; Wang, J.; Zhang, Q. A broadband low profile microstrip filter-antenna with an omni-directional pattern. In Proceedings of the 2014 3rd Asia-Pacific Conference on Antennas and Propagation, Harbin, China, 26–29 July 2014; pp. 580–582.

63. Lin, C.; Chung, S. A Compact Filtering Microstrip Antenna with Quasi-Elliptic Broadside Antenna Gain Response. *IEEE Antennas Wirel. Propag. Lett.* **2011**, *10*, 381–384.

64. Wu, W.; Ma, B.; Wang, J.; Wang, C. Design of a microstrip antenna with filtering characteristics for wireless communication systems. In Proceedings of the 2017 Sixth Asia-Pacific Conference on Antennas and Propagation (APCAP), Xi'an, China, 16–19 October 2017; pp. 1–3.

65. Al-Yasir, Y.I.A.; Parchin, N.O.; Abd-Alhameed, R.A. New High-Gain Differential-Fed Dual-Polarized Filtering Microstrip Antenna for 5G Applications. In Proceedings of the 14th European Conference on Antennas and Propagation (EuCAP), Copenhagen, Denmark, 15–20 March 2020; pp. 1–5.

66. Wu, W.; Yin, Y.; Zuo, S.; Zhang, Z.; Xie, J. A New Compact Filter-Antenna for Modern Wireless Communication Systems. *IEEE Antennas Wirel. Propag. Lett.* **2011**, *10*, 1131–1134.

67. Ohira, M.; Ma, Z. An efficient design method of microstrip filtering antenna suitable for circuit synthesis theory of microwave band-pass filters. In Proceedings of the 2015 International Symposium on Antennas and Propagation (ISAP), Hobart, Australia, 9–12 November 2015; pp. 1–4.

68. Al-Yasir, Y.I.A.; Alhamadani, H.A.; Kadhim, A.S.; Ojaroudi Parchin, N.; Saleh, A.L.; Elfergani, I.T.E.; Rodriguez, J.; Abd-Alhameed, R.A. Design of a Wide-Band Microstrip Filtering Antenna with Modified Shaped Slots and SIR Structure. *Inventions* **2020**, *5*, 11. [CrossRef]

69. Abdel-Jabbar, H.; Kadhim, A.S.; Saleh, A.L.; Al-Yasir, Y.I.A.; Parchin, N.O.; Abd-Alhameed, R.A. Design and optimization of microstrip filtering antenna with modified shaped slots and SIR filter to improve the impedance bandwidth. *TELKOMNIKA* **2020**, *18*, 515–545. [CrossRef]

70. Mohammed, K.A.; Alhamadani, A.; Al-Yasir, Y.I.A.; Saleh, A.L.; Parchin, N.O.; Abd-Alhameed, R. Study on the effect of the substrate material type and thickness on the performance of the filtering antenna design. *TELKOMNIKA* **2020**, *18*, 72–79.

71. Yang, D.; Zhai, H.; Guo, C.; Li, H. A Compact Single-Layer Wideband Microstrip Antenna with Filtering Performance. *IEEE Antennas Wirel. Propag. Lett.* **2020**, *19*, 801–805. [CrossRef]

72. Tawk, Y.; Costantine, J.; Christodoulou, C.G. A varactor-based reconfigurable filtenna, *IEEE Antennas Wirel. Propag. Lett.* **2012**, *11*, 716–719.

73. Soltanpour, M.; Fakharian, M.M. Compact filtering slot antenna with frequency agility for Wi-Fi/LTE mobile applications. *Electron. Lett.* **2016**, *52*, 491–492. [CrossRef]

74. Augustin, G.; Chacko, B.P.; Denidni, T.A. Electronically reconfigurable uni-planar antenna for cognitive radio applications. *Microw. Antennas Propag.* **2014**, *8*, 367–376. [CrossRef]

75. Deng, J.; Hou, S.; Zhao, L.; Guo, L. Wideband-to-narrowband tunable monopole antenna with integrated band-pass filters for UWB/WLAN applications. *IEEE Antennas Wirel. Propag. Lett.* **2017**, *16*, 2734–2737. [CrossRef]

76. Kingsly, S.; Thangarasu, D.; Alsath, M.G.N.; Thipparaju, R.R.; Palaniswamy, S.K.; Sambandam, P. Multiband Reconfigurable Filtering Monopole Antenna for Cognitive Radio Applications. *IEEE Antennas Wirel. Propag. Lett.* **2018**, *17*, 1416–1420. [CrossRef]

77. Hu, P.F.; Pan, Y.M.; Zhang, X.Y.; Hu, B. A Filtering Patch Antenna with Reconfigurable Frequency and Bandwidth Using F-Shaped Probe. *IEEE Trans. Antennas Propag.* **2019**, *67*, 121–130. [CrossRef]

78. Rodrigues, L.; Varum, T.; Matos, J.N. The Application of Reconfigurable Filtennas in Mobile Satellite Terminals. *IEEE Access* **2020**, *8*, 77179–77187. [CrossRef]

79. Mahmoud, K.R.; Montaser, A.M. Design of Compact mm-wave Tunable Filtenna Using Capacitor Loaded Trapezoid Slots in Ground Plane for 5G Router Applications. *IEEE Access* **2020**, *8*, 27715–27723. [CrossRef]

80. Bi, X.; Zhang, X.; Wong, S.; Guo, S.; Yuan, T. Design of Notched-Wideband Bandpass Filters with Reconfigurable Bandwidth Based on Terminated Cross-Shaped Resonators. *IEEE Access* **2020**, *8*, 37416–37427. [CrossRef]

81. Qin, P.-Y.; Wei, F.; Guo, Y.J. A wideband-to-narrowband tunable antenna using a reconfigurable filter. *IEEE Trans. Antennas Propag.* **2015**, *63*, 2282–2285. [CrossRef]

82. Tang, M.-C.; Wen, Z.; Wang, H.; Li, M.; Ziolkowski, R.W. Compact, frequency-reconfigurable filtenna with sharply defined wideband and continuously tunable narrowband states. *IEEE Trans. Antennas Propag.* **2017**, *65*, 5026–5034. [CrossRef]

83. Horestani, A.K.; Shaterian, Z.; Naqui, J.; Martín, F.; Fumeaux, C. Reconfigurable and tunable S-shaped split-ring resonators and application in band-notched UWB antennas. *IEEE Trans. Antennas Propag.* **2016**, *64*, 3766–3776. [CrossRef]

84. Malakooti, S.-A.; Mousavi, S.M.H.; Fumeaux, C. Tunable bandpassto-bandstop quasi-Yagi–Uda antenna with sum and difference radiation patterns. *IEEE Trans. Antennas Propag.* **2019**, *67*, 2260–2271. [CrossRef]

85. Yang, X.; Lin, H.; Gu, H.; Ge, L.; Zeng, X. Broadband pattern diversity patch antenna with switchable feeding network. *IEEE Access* **2018**, *6*, 69612–69619. [CrossRef]

86. Yassin, M.E.; Mohamed, H.A.; Abdallah, E.A.F.; El-Hennawy, H.S. Circularly Polarized Wideband-to-Narrowband Switchable Antenna. *IEEE Access* **2019**, *7*, 36010–36018. [CrossRef]

87. Alhegazi, A.; Zakaria, Z.; Shairi, N.; Salleh, A.; Qasem, S. Review of Recent Developments in Filtering-Antennas. *Int. J. Commun. Antenna Propag. (IRECAP)* **2016**, *6*, 125–131. [CrossRef]

88. Ahmed, S.; Geok, T.; Alias, M.; Kamaruddin, M. A Survey on Recent Developments in Filtering Antenna Technology. *Int. J. Commun. Antenna Propag. (IRECAP)* **2018**, *8*, 374–384. [CrossRef]

89. Sam, W.Y.; Zakaria, Z. A Review on Reconfigurable Integrated Filter and Antenna. *Prog. Electromagn. Res. B* **2015**, *63*, 263–273. [CrossRef]

90. Rohde, U.L.; Newkirk, D.P. *RF/Microwave Circuit Design for Wireless Applications*; Wiley: New York, NY, USA, 2000.

91. Cetiner, B.A.; Crusats, G.R.; Jofre, L.; Biyikli, N. RF MEMS integrated frequency reconfigurable annular slot antenna. *IEEE Trans. Antennas Propag.* **2010**, *58*, 626–632. [CrossRef]

92. Chen, R.-H.; Row, J.-S. Sing-fed microstrip patch antenna with switchable polarization. *IEEE Trans. Antennas Propag.* **2008**, *56*, 922–926. [CrossRef]

93. Behdad, N.; Sarabandi, K. A varactor-tuned dual-band slot antenna. *IEEE Trans. Antennas Propag.* **2006**, *54*, 401–408. [CrossRef]

94. Al-Yasir, Y.; Abd-Alhameed, R.A.; Noras, J.M.; Abdulkhaleq, A.M.; Ojaroudi, N. Design of Very Compact Combline Band-Pass Filter for 5G Applications. In Proceedings of the Loughborough Antennas & Propagation Conference (LAPC), Loughborough, UK, 12–13 November 2018; pp. 1–4.

95. Zhao, D.; Lan, L.; Han, Y.; Liang, F.; Zhang, Q.; Wang, B.Z. Optically controlled reconfigurable band-notched UWB antenna for cognitive radio applications. *IEEE Photon. Technol. Lett.* **2014**, *26*, 2173–2176. [CrossRef]

96. Parchin, N.O.; Al-Yasir, Y.I.A.; Abd-Alhameed, R.A. *Microwave/RF Components for 5G Front-End Systems*; AVID SCIENCE: Telangana, India, 2020.

97. Chen, Z.; Wong, H.; Kelly, J. A polarization-reconfigurable glass dielectric resonator antenna using liquid metal. *IEEE Trans. Antennas Propag.* **2019**, *67*, 3427–3432. [CrossRef]

98. Al-Yasir, Y.I.A.; Tu, Y.; Parchin, N.O.; Elfergani, I.; Abd-Alhameed, R.; Rodriguez, J.; Noras, J. Mixed-coupling multi-function quint-wideband asymmetric stepped impedance resonator filter. *Microw. Opt. Tech. Lett.* **2019**, *61*, 1181–1184. [CrossRef]

99. Zheng, S.H.; Liu, X.; Tentzeris, M.M. Optically controlled reconfigurable band-notched UWB antenna for cognitive radio systems. *Electron. Lett.* **2014**, *50*, 1502–1504. [CrossRef]

100. Ullah, U.; Koziel, S.; Mabrouk, I.B. Rapid Redesign and Bandwidth/Size Tradeoffs for Compact Wideband Circular Polarization Antennas Using Inverse Surrogates and Fast EM-Based Parameter Tuning. *IEEE Trans. Antennas Propag.* **2020**, *68*, 81–89. [CrossRef]

101. Row, J.; Kuo, L. Pattern-Reconfigurable Array Based on a Circularly Polarized Antenna with Broadband Operation and High Front-to-Back Ratio. *IEEE Trans. Antennas Propag.* **2020**, *68*, 4109–4113. [CrossRef]

102. Wu, Z.; Tang, M.; Li, M.; Ziolkowski, R.W. Ultralow-Profile, Electrically Small, Pattern-Reconfigurable Metamaterial-Inspired Huygens Dipole Antenna. *IEEE Trans. Antennas Propag.* **2020**, *68*, 1238–1248. [CrossRef]

103. Li, Y.; Zhao, Z.; Tang, Z.; Yin, Y. Differentially Fed, Dual-Band Dual-Polarized Filtering Antenna with High Selectivity for 5G Sub-6 GHz Base Station Applications. *IEEE Trans. Antennas Propag.* **2020**, *68*, 3231–3236. [CrossRef]

104. Hao, Z.; Hong, J. Quasi-Elliptic UWB Bandpass Filter Using Multilayer Liquid Crystal Polymer Technology. *IEEE Microw. Wirel. Compon. Lett.* **2010**, *20*, 202–204. [CrossRef]

105. Zhou, P.; Li, B.; Lu, H.; Shi, Y.; Tang, W. Novel Ultrawideband and Multimode LTCC Common-Mode Filter Based on the Dual Vertical Coupling Paths. *IEEE Trans. Compon. Packag. Manuf. Technol.* **2019**, *9*, 1345–1353. [CrossRef]

106. Shen, G.; Che, W. Compact Ku-band LTCC bandpass filter using folded dual-composite right- and left-handed resonators. *Electron. Lett.* **2020**, *56*, 17–19. [CrossRef]

107. Shen, Z.; Xu, K.; Shi, J. Compact single-layer bandwidth-enhanced balanced bandpass filter using half-mode substrate-integrated waveguide. *Electron. Lett.* **2019**, *55*, 697–699. [CrossRef]

108. Sans, M.; Selga, J.; Vélez, P.; Bonache, J.; Martín, F.; Rodríguez, A.; Boria, V.E. Optimized wideband differential-mode bandpass filters with broad stopband and common-mode suppression based on multi-section stepped impedance resonators and interdigital capacitors. In Proceedings of the IEEE MTT-S International Conference on Numerical Electromagnetic and Multiphysics Modeling and Optimization for RF, Microwave, and Terahertz Applications (NEMO), Seville, Spain, 17–19 May 2017; pp. 10–12.

109. Jamshidi, M.; Lalbakhsh, A.; Lotfi, S.; Siahkamari, H.; Mohamadzade, B.; Jalilian, J. A neuro-based approach to designing a Wilkinson power divider. *Int. J. RF Microw. Comput. Aided Eng.* **2020**, *30*, 1–10. [CrossRef]

110. Mohammed, H.J.; Abdulsalam, F.; Abdulla, A.S.; Ali, R.S.; Abd-Alhameed, R.A.; Noras, J.M.; Abdulraheem, Y.I.; Ali, A.; Rodriguez, J.; Abdalla, A.M. Evaluation of genetic algorithms, particle swarm optimisation, and firefly algorithms in antenna design. In Proceedings of the 13th International Conference on Synthesis, Modeling, Analysis and Simulation Methods and Applications to Circuit Design (SMACD), Lisbon, Portugal, 27–30 June 2016; pp. 1–4.

111. Mohammed, H.J.; Abdullah, A.S.; Ali, R.S.; Abd-Alhameed, R.A.; Abdulraheem, Y.I.; Noras, J.M. Design of a uniplanar printed triple band-rejected ultra-wideband antenna using particle swarm optimisation and the firefly algorithm. *IET Microw. Antennas Propag.* **2016**, *10*, 31–37. [CrossRef]

Article

Modified U-Shaped Resonator as Decoupling Structure in MIMO Antenna

Amjad Iqbal [1,2]**, Ahsan Altaf** [3]**, Mujeeb Abdullah** [4]**, Mohammad Alibakhshikenari** [5]**, Ernesto Limiti** [5] **and Sunghwan Kim** [6,*]

[1] Centre for Wireless Technology, Faculty of Engineering, Multimedia University, Cyberjaya 63100, Malaysia; amjad730@gmail.com

[2] Electrical Engineering Department, CECOS University of IT and Emerging Sciences, Peshawar 25000, Pakistan

[3] Department of Electrical Engineering, Istanbul Medipol University, Istanbul 34083, Turkey; aaltaf@st.medipol.edu.tr

[4] Department of Computer Science, Bacha Khan University, Charsadda 24420, Pakistan; mujeeb.abdullah@gmail.com

[5] Electronic Engineering Department, University of Rome "Tor Vergata", Via del Politecnico 1, 00133 Rome, Italy; alibakhshikenari@ing.uniroma2.it (M.A.); limiti@ing.uniroma2.it (E.L.)

[6] School of Electrical Engineering, University of Ulsan, Ulsan 44610, Korea

* Correspondence: sungkim@ulsan.ac.kr; Tel.: +82-52-259-1401

Received: 21 July 2020; Accepted: 14 August 2020; Published: 16 August 2020

Abstract: This paper presents an isolation enhancement of two closely packed multiple-input multiple-output (MIMO) antenna system using a modified U-shaped resonator. The modified U-shaped resonator is placed between two closely packed radiating elements resonating at 5.4 GHz with an edge to edge separation distance of 5.82 mm ($\lambda_o/10$). Through careful adjustment of parametric modelling, the isolation level of -23 dB among the densely packed elements is achieved. The coupling behaviour of the MIMO elements is analysed by accurately designing the equivalent circuit model in each step. The antenna performance is realized in the presence and absence of decoupling structure, and the results shows negligible effects on the antenna performance apart from mutual coupling. The simple assembly of the proposed modified U-shaped isolating structure makes it useful for several linked applications. The proposed decoupling structure is compact in nature, suppress the undesirable coupling generated by surface wave and nearby fields, and is easy to fabricate.

Keywords: isolation enhancement; surface waves; gain; circuit model; line resonators; MIMO antennas

1. Introduction

Multiple-input multiple-output (MIMO) antennas have been given the centre of attention due to their higher performance characteristics. Planar antennas (PA) are unique minute elements, when assembled in multiple elements array form can deliver beam forming, pattern and spatial diversity, focused directivity, and higher gain characteristics [1–3]. In the MIMO arrays system when two or more than two radiating elements are excited simultaneously for better performance, their proximity in near fields and surface wave currents give rise to a coupling, which can alter MIMO array results consequentially, leading to system failure and undesired results.

Several techniques have been reported in the literature to reduce mutual coupling. The isolation techniques include the insertion of slots or ground irregularities [4–7], uniplanar, mushroom type, and various shapes of electromagnetic bandgap structures (EBG) [8–12], split-ring resonators (SRRs) [13–15], metamaterials [16] and meander line resonators [17–19]. Not only these techniques but several antenna configurations such as superstrate configurations of photonic bandgap structures [20] and orientations have been reported to be useful in minimizing coupling effects. Applying a 90-degree phase difference have resulted in enhanced isolation among driven and parasitic patch [21–24]. In [14], an efficiently folded split ring resonator (FSRR) is presented, minimizing near field coupling effects up to great extent, however, with the insertion of the proposed structure, an increase in backward radiation is seen which resulted in a reduced gain, front to back ratio and radiation efficiency. A novel EBG structure is presented in [25], consisting of an in-house package of four EBG structures. The insertion of the periodic structure at one half wavelength frequency reduced coupling up to 22.7 dB without disrupting gain; however, it has lower antenna efficiency with a slight increase in the back lobe radiation. The use of EBG has shown active capability in suppressing surface waves in MIMO systems [26] but they usually require larger distances (up to half wavelength in free space) and give rise to undesirable radio leakage complications. Similarly, a dual negative photonic bandgap structure is proposed in [27] in which the coupling energy is confined by superstrate orientation of the proposed structure which reportedly reduced the coupling up to 23 dB, however, the design and fabrication process increases complexity, stability issues and overall cost of the system. The DGS presented in [28,29] show a significant decrease of up to 40 dB in isolation level however with the drawback of more back lobe radiation patterns and reduced gains. A modified serpentine structure is used between two elements of MIMO antenna to reduce the coupling up to 10–34 dB in the operating bandwidth [30].

In this paper, a simple modified U-shaped resonator is presented. A two-element MIMO system is designed at a close distance of 5.82 mm (edge to edge). The isolation of more than 20 dB is achieved after insertion of the proposed isolating structure. Furthermore, the input impedance of the equivalent circuit of the model with and without the proposed model matches well with the Electromagnetic (EM) model. The proposed line resonator inhabits a very low occupying area and so is simple and easy to implement. The proposed antenna has advantages over existing antennas in terms of small size of decoupling structure, lower edge to edge distance, high FTBR value, lower ECC and CCL values. The high FTBR value of the antenna make this structure suitable for beam scanning and other applications where high directivity is required. This paper is organized as following: Section 1 covers a detail literature review on isolation enhancement among MIMO antenna elements. Section 2 covers the proposed antenna design with and without isolation structure. In the Results and Discussion, scattering parameters with surface current distribution and radiation patterns, along with other MIMO performance characteristics are presented, which is followed by the Conclusion.

2. Design Methodology

The proposed MIMO antenna is printed on an FR4 substrate with relative permittivity of 4.4 and a thickness of 1.6 mm. A unit antenna is designed by using standard equations [31] at a resonance frequency of 5.4 GHz and is transformed into MIMO separated at less than half-wavelength apart. The proposed antenna is excited through a 50 Ω transmission line having 3.1 mm width. Figure 1 shows geometry of the proposed MIMO Antenna. The design dimensions of proposed antenna are: a_p = 16.8 mm, b_p = 12.91 mm, gap = 5.82 mm, W_s = 3.1 mm, g = 0.7 mm, and lg = 4.5 mm.

The radiation resistance of an antenna plays a vital role in the performance of the antenna. It is one of the key performance elements of the antenna. The study of radiation resistance of the antenna as a function of inset feeding width (g) and length (lg) is presented in Figure 2. It is observed that by increasing the feeding width (g), the radiation resistance decreased. Similarly, radiation resistance decreased as we increased the length (lg) of the inset feed.

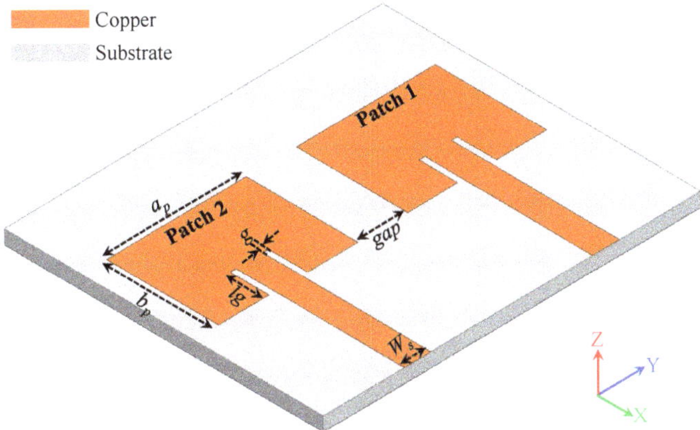

Figure 1. Geometry of the initially designed MIMO antenna (a_p = 16.8, b_p = 12.91, gap = 5.82, W_s = 3.1, g = 0.7, and lg = 4.5 [unit = mm]).

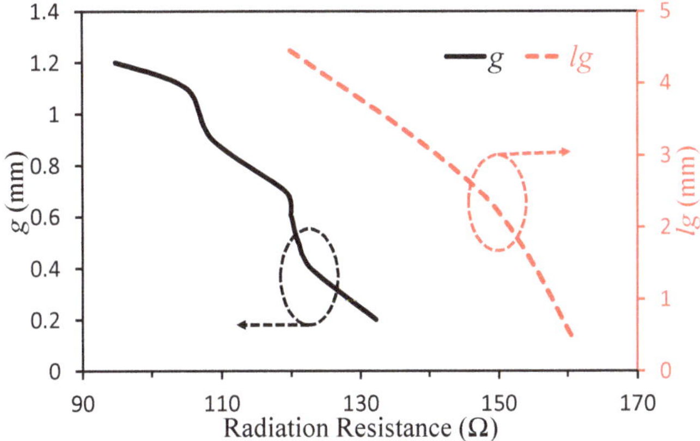

Figure 2. Radiation resistance of the antenna with varying g (when lg = 4.5 mm) and lg (when g = 0.7 mm).

2.1. Unit Antenna

The electrical model is helpful to understand the operating of the antenna in terms of equivalent lump elements for better illustration. From the open literature presented in [28,32,33], a resonator can be modelled as a parallel RLC circuit. Using the same theory, a single element equivalent circuit model is designed. The patch antenna is approximated by an RLC circuit, where R is the radiation resistance of the radiating mode of the antenna, and L and C describe the resonant circuit that are responsible for the desired resonant frequency. The circuit model of the antenna system based on the circuit theory is shown in Figure 3a. The transmission line is modelled as an impedance transformer with the coupling ratio of X:1 in the circuit model. The equivalent circuit model was optimized in Keysight Advance Design System (ADS). The values of each component in the circuit model are depicted in the Figure 3a. The results obtained from the circuit model are verified with the full-wave simulation analysis, as shown in Figure 3b. We observed a close match between the input impedance result of both circuit.

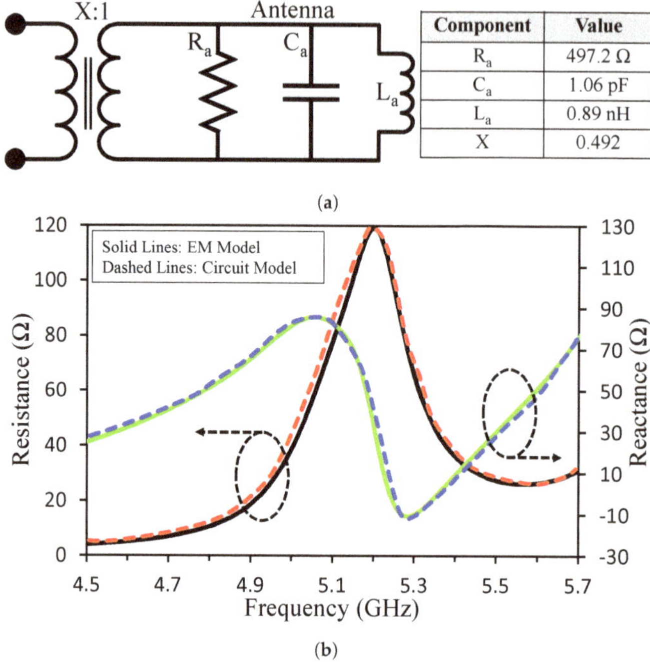

Component	Value
R_a	497.2 Ω
C_a	1.06 pF
L_a	0.89 nH
X	0.492

(a)

(b)

Figure 3. (**a**) Equivalent circuit model of the single-unit antenna, and (**b**) input impedance (resistance and reactance) comparison of the EM model (solid lines) and circuit model (dashed lines).

2.2. MIMO Antenna without Decoupling Structure

In this section, the coupling behavior of the antenna elements is analyzed in detail with the exact electrical model description. The main challenge for the MIMO antenna is port isolation or low mutual coupling due to the integration of multiple radiating elements on the small footprint area of the printed circuit board. The two main phenomena namely surface wave propagation inside the substrate and space wave propagation related to near-field/reactive coupling contributes to a large extent in the coupled patch antennas [34]. The after-mentioned limiting factors for MIMO antenna ports isolation can be easily understand via coupled-resonator theory with equivalent resonant circuits.

As mentioned above, the resonators can be represented in the RLC parallel circuit. The circuit model of a two-element MIMO antenna without a decoupling network is presented in Figure 4a. As discussed earlier, the antenna patches and the two transmission lines are modelled using RLC circuits and impedance transformers, respectively. The details of each component are depicted in the table. The coupling depends on the distance between the antenna elements. Since the two antenna elements are near, they are coupled, and the coupling is represented by coupling coefficient (K_{12}) which is calculated as given in [35]. Strong reactive coupling of around -9 dB within the desired bandwidth is observed. An excellent agreement between the computed circuit model results with the full-wave analysis results is shown in Figure 4b.

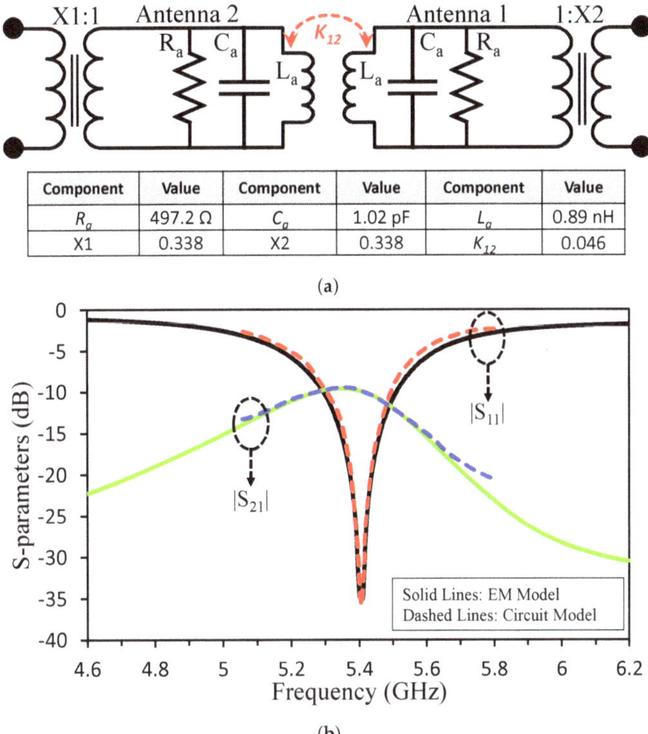

Component	Value	Component	Value	Component	Value
R_a	497.2 Ω	C_a	1.02 pF	L_a	0.89 nH
X1	0.338	X2	0.338	K_{12}	0.046

(a)

(b)

Figure 4. (**a**) Equivalent circuit model of the two-element MIMO antenna without decoupling structure, and (**b**) S-parameters comparison of the EM model (solid lines) and circuit model (dashed lines).

2.3. MIMO Antenna with Decoupling Structure

A novel shaped decoupling structure is used to eliminate the coupling between the MIMO antennas as shown in Figure 5. The dimensions of the decoupling structure are as c = 5 mm, d = 22 mm, e = 20 mm, and f = 2 mm. After the insertion of the decoupling structure, the isolation of −23 dB for the MIMO antenna is accomplished.

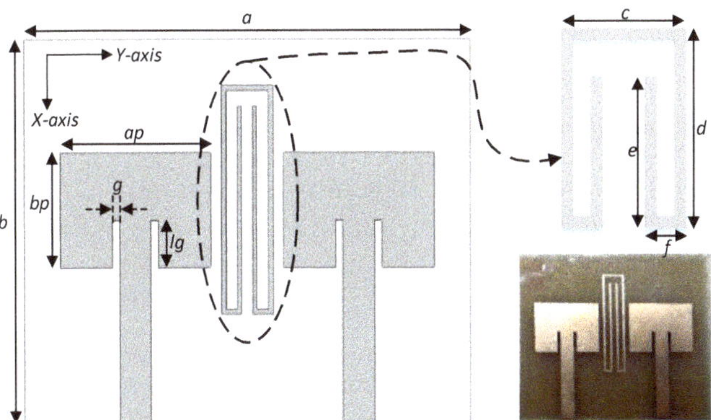

Figure 5. Geometry of the proposed MIMO antenna with decoupling structure (a_p = 16.8, b_p = 12.91, W_s = 3.1, g = 0.7, and lg = 7, a = 44, b = 37, c = 5, d = 22, e = 20, f = 2 [unit = mm]) and fabricated prototype.

The circuit model of a two-element MIMO antenna with a decoupling network is presented in Figure 6a. A parallel combination of RLC is used as a decoupling network, and impedance transformers and RLC circuits are used to model two transmission lines and the antenna patches respectively. The values of each component in the design are depicted in the table. The coupling between the antenna elements is reduced by generating a modified U-shaped resonator, which in turn suppress the undesirable coupling in the frequency range. By the introduction of a modified U-shaped resonator, a strong coupling between the antenna elements is suppressed and a high reduction of coupling around −23 dB is achieved. The validation of the circuit model with the EM analysis is shown in Figure 6b.

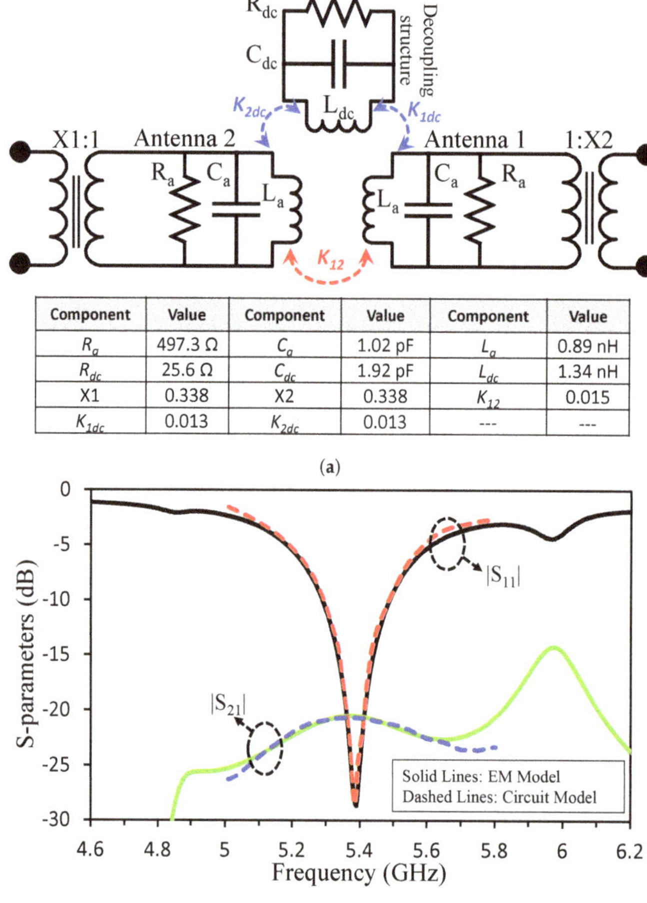

Component	Value	Component	Value	Component	Value
R_a	497.3 Ω	C_a	1.02 pF	L_a	0.89 nH
R_{dc}	25.6 Ω	C_{dc}	1.92 pF	L_{dc}	1.34 nH
X1	0.338	X2	0.338	K_{12}	0.015
K_{1dc}	0.013	K_{2dc}	0.013	---	---

(**a**)

(**b**)

Figure 6. (**a**) Equivalent circuit model of the two-element MIMO antenna with decoupling structure, and (**b**) S-parameters comparison of the EM model (solid lines) and circuit model (dashed lines).

We performed many simulations to show the impact of decoupling structure's dimensions on the reflection and mutual coupling of the antenna. Initially, parameter c was analysed between 4 and 5.5 mm. The parametric analysis of c is shown in Figure 7a. The mutual coupling between the MIMO elements increased when the value of c was reduced or increased from 5 mm. The resonant frequency of the antenna remained unchanged. If we compare the results with circuit model of the decoupling structure, it can be observed that parameter c mainly contribute in K_{1dc} and K_{2dc} and has small impact on R_{dc}, C_{dc} and L_{dc}. The parametric analysis of d is shown in Figure 7b. We see that the coupling value is inconsistent with variation in d. Analysing the results in context of circuit model, it can be noted that parameter d has impact on K_{1dc} and K_{2dc} as well as on R_{dc}, C_{dc} and L_{dc}.

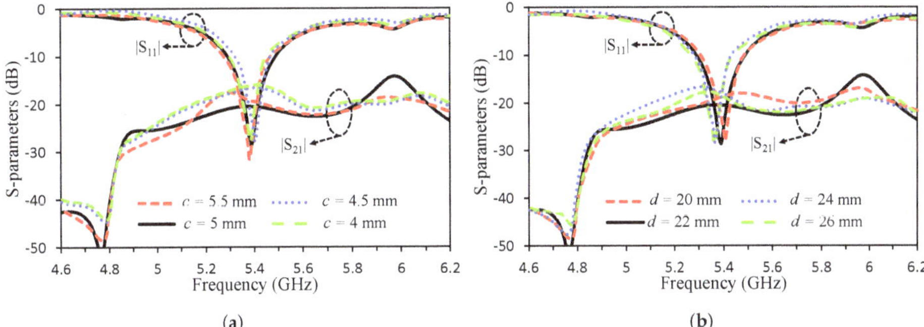

Figure 7. Parametric analysis of the decoupling structure by varying (**a**) c and (**b**) d.

3. MIMO Parameters

3.1. Envelop Correlation Coefficient (ECC) and Diversity Gain (DG)

Envelop Correlation Coefficient (ECC) and Diversity Gain (DG) are the important MIMO antenna parameters. ECC is the measure of how well antennas are correlated to each other in terms of performance characteristics like return loss and far-field results while diversity gain is the selection of strongest signal for N number of signals. The ECC can be calculated using S-parameters (see Equation (1)) or far-field patterns (see Equation (2)). We calculated the ECC using far field. The ECC and DG are calculated using the following formula given in [11].

$$ECC = \frac{|S_{11}^* S_{12} + S_{22}^* S_{21}|^2}{[1 - (|S_{11}|^2 + |S_{12}|^2)][1 - (|S_{22}|^2 + |S_{21}|^2)]} \tag{1}$$

$$ECC = \frac{|\iint_{4\pi} (\vec{B}_i(\theta,\phi)) \times (\vec{B}_j(\theta,\phi))\, d\Omega|^2}{\iint_{4\pi} |(\vec{B}_i(\theta,\phi))|^2\, d\Omega \iint_{4\pi} |(\vec{B}_j(\theta,\phi))|^2\, d\Omega} \tag{2}$$

where S_{11}/S_{22} and S_{21}/S_{12} are the reflection and transmission coefficient of the antenna. $\vec{B}_i(\theta,\phi)$ is the three dimensional radiation pattern upon excitation of the ith antenna and $\vec{B}_j(\theta,\phi)$ is the three dimensional radiation pattern upon excitation of the jth antenna. Ω represents the solid angle.

$$DG = 10\sqrt{1 - (ECC)^2} \tag{3}$$

The ECC and DG of the MIMO Antenna after the insertion of the proposed isolating structure is given in Figure 8. From the figure, it can be seen that ECC based on far-field results is less than 0.1 and DG is 9.95 dBi at 5.4 GHz with greater than 9.94 dB value for a complete band of interest.

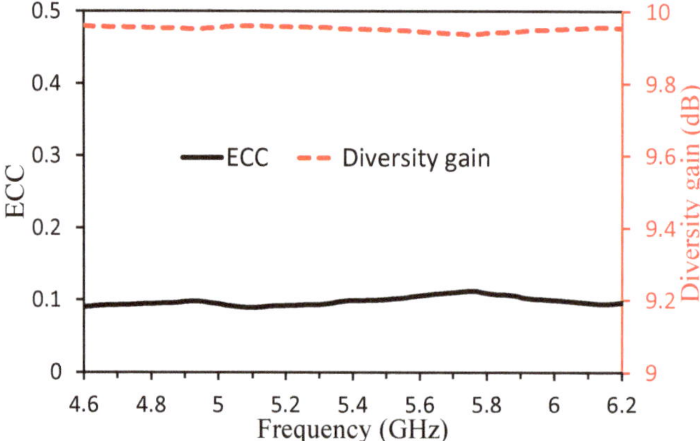

Figure 8. ECC and diversity gain of the antenna.

3.2. Channel Capacity Loss (CCL)

The channel capacity loss (CCL) is an important parameter of MIMO antenna. From Figure 9, it can be seen from simulated and measured results that CCL value is equal to 0.07 at 5.4 GHz and less than 0.09 in the whole operating band.

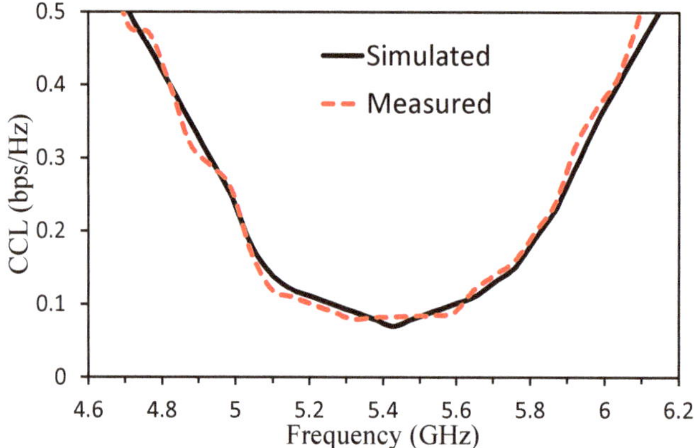

Figure 9. Simulated and measured CCL of the antenna in the presence of decoupling structure.

4. Results and Discussions

Surface currents distribution explains the performance analysis of isolating structure over the desired frequency band. Figure 10 shows the current distribution of the MIMO antenna with and without decoupling structure. From Figure 10, it is clear that without the proposed decoupling structure the currents from one antenna upon excitation is seen with high concentrations to another, consequentially leading in high coupling levels whereas with the insertion of decoupling structure, the concentration of currents is focused on the edges of the decoupling structure, resulting in higher isolation.

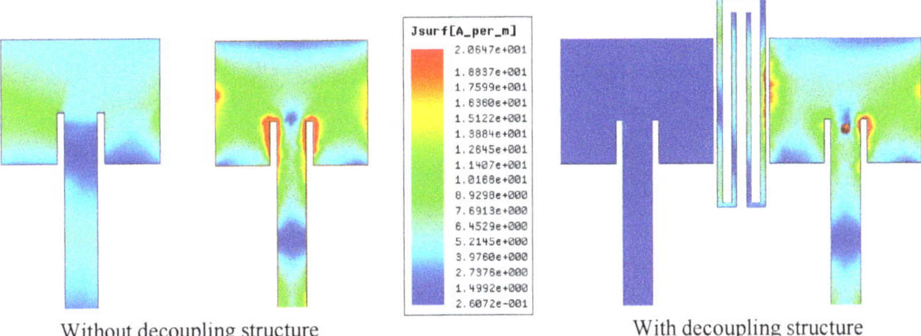

Figure 10. Surface current distribution of the antenna at 5.4 GHz in the presence and absence of decoupling structure.

The proposed antenna was designed and simulated in ANSYS HFSS and was fabricated and tested. The Reflection coefficient and isolation level parameters of the two-port MIMO antenna were measured using a Vector Network Analyzer (VNA) and the radiation patterns were measured in an anechoic chamber. The VNA was calibrated using calibration kit and the two antennas were connected with the two ports for S-parameter measurements. The simulated and measured results were in slightly shift which can be due to cable and environmental losses and errors. Figure 11 shows the simulated and measured S parameters of the MIMO antenna with a decoupling structure as well as simulated S parameters without decoupling structure. It can be seen from the figure that after the insertion of decoupling structure the isolation improvement of more than 14 dB is achieved. We can observe that the reflection coefficient of the MIMO antenna with and without decoupling element is same.

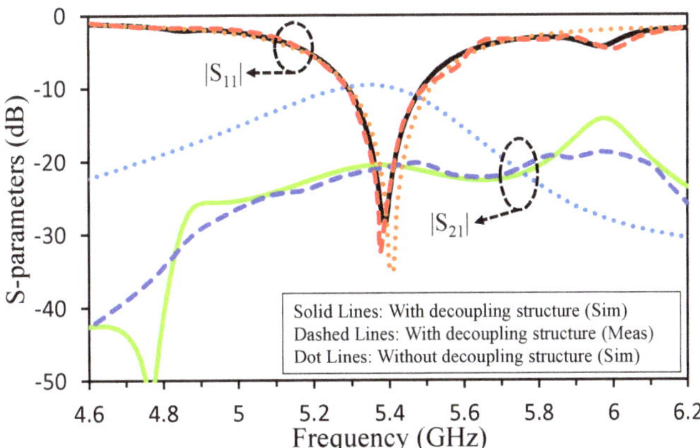

Figure 11. Simulated (with and without decoupling structure) and measured (with decoupling structure) S-parameters of the antenna.

The antenna efficiency and maximum gain over frequency are given in Figure 12. The gain of two-port MIMO antenna with and without decoupling structure is 4.24 dBi and 4.25 dBi, respectively at 5.4 GHz and greater than 4.24 in the whole operating band. Similarly, the efficiency of the MIMO antenna is 81.2% at 5.4 GHz and >80.8% in the whole operating band without decoupling structure. With decoupling structure, the efficiency was 82.4% at 5.4 GHz and >81.1% in the whole operating band. The results shows that the efficiency and gain of the antenna parameters have not been affected after the isolation enhancement.

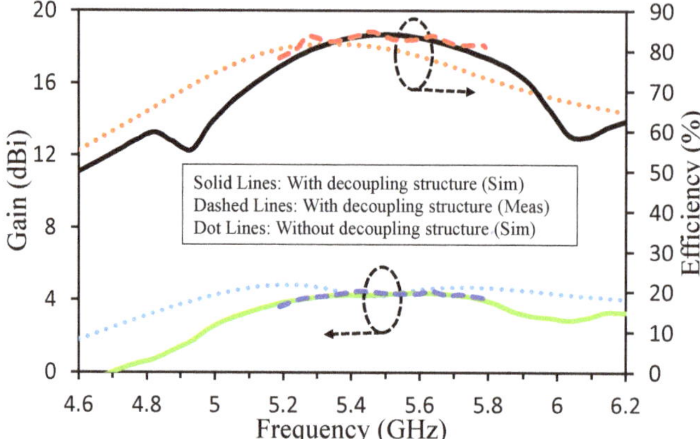

Figure 12. Simulated (with and without decoupling structure) and measured (with decoupling structure) gain and efficiency of the antenna.

The radiation patterns of the antenna in two main planes were measured inside an anechoic chamber. One port of the antenna was connected with spectrum analyser and the second port was terminated with a 50 Ω load. A high gain horn antenna was used as a transmitter. The simulated and measured radiation patterns of the proposed MIMO antenna are shown in Figure 13. The Elevation and Azimuth (*E* and *H*) plane patterns are slightly misaligned as compared to simulated results which can be justified due to environmental losses and errors. However, it can be observed that the radiation patterns both in *E* and *H* planes have remained the same before and after the insertion of the decoupling structure. The proposed system has uni-directional radiation pattern with high front to back ratio of 22.1 dB at 5.4 GHz.

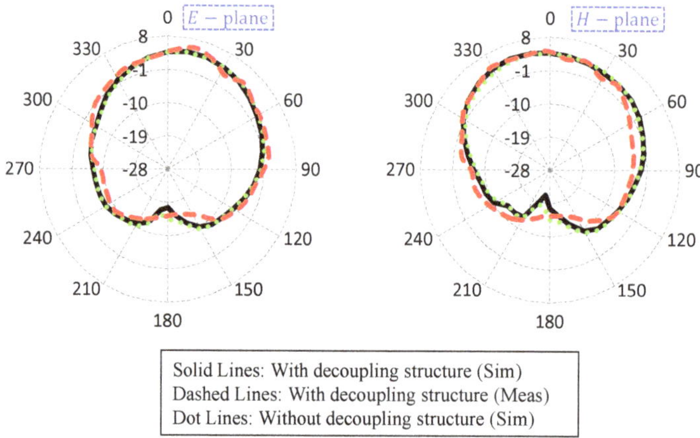

Figure 13. Simulated (with and without decoupling structure) and measured (with decoupling structure) radiation pattern (*E*- and *H*-plane) of the antenna at 5.4 GHz.

A comparison between our proposed antenna and other antennas is listed in Table 1. It can be observed that the proposed antenna has advantages over existing antennas in terms of small size of decoupling structure, lower edge to edge distance, high FTBR value, lower ECC and CCL values. The high FTBR value of the antenna make this structure suitable for beam scanning and other applications where high directivity is required.

Table 1. Comparison table (where $\lambda_\circ$ is the wavelength at centre frequency).

Ref.	Technique	Centre Frequency (GHz)	Edge to Edge Distance	Isolation Improvement	FTBR (dB)	ECC	CCL (bps/Hz)
[7]	Slotted ground	5.8	$0.33\lambda_\circ$	40	NA	NA	NA
[9]	EBG	7.5	NA	4	NA	NA	NA
[10]	UC-EBG	5.56	$0.5\lambda_\circ$	10	NA	NA	NA
[11]	Metamaterial	5.8	$0.135\lambda_\circ$	9	NA	<0.1	<0.05
[17]	I-shaped resonator	2.8	$0.056\lambda_\circ$	8–10	NA	NA	NA
[26]	EBG	6	$0.5\lambda_\circ$	8	NA	<0.01	NA
[30]	Serpentine structure	2.45	$0.05\lambda_\circ$	10–34	NA	<0.007	NA
[34]	metamaterial polarization-rotator	60	NA	16	NA	$<0.1 \times 10^{-6}$	NA
This Work	**U-Shaped resonator**	5.4	$0.1\lambda_\circ$	14	22.1	<0.1	0.07

5. Conclusions

We presented a simple and efficient technique to reduce the mutual coupling between nearly packed MIMO elements. We used a modified U-shaped resonator as decoupling structure between the MIMO elements to reduce the undesired associated coupling between them. The MIMO elements were kept at a distance of $\lambda_\circ/10$ (edge to edge) and the coupling suppression of 14 dB was achieved. The coupling behaviour of the MIMO elements was studied using coupled resonator theory. The proposed system has uni-directional radiation pattern with high front to back ratio of 22.1 dB at 5.4 GHz.

Author Contributions: Design and concept, A.I.; methodology, A.I., A.A. and M.A. (Mujeeb Abdullah); investigation, A.I.; resources, A.I. and S.K.; writing—original draft preparation, A.I., A.A. and M.A. (Mujeeb Abdullah); writing—review and editing, A.I., A.A., M.A. (Mujeeb Abdullah), M.A. (Mohammad Alibakhshikenari), E.L. and S.K.; validation, A.I., A.A., M.A. (Mujeeb Abdullah), M.A. (Mohammad Alibakhshikenari), E.L. and S.K.; supervision, M.A. (Mohammad Alibakhshikenari), E.L. and S.K.; project administration, M.A. (Mohammad Alibakhshikenari), E.L. and S.K. All authors have read and agreed to the published version of the manuscript.

Funding: Following are results of a study on the "Leaders in INdustry-university Cooperation +" Project, supported by the Ministry of Education and National Research Foundation of Korea.

Conflicts of Interest: The authors declare no conflict of interest.

References

1. Abdullah, M.; Kiani, S.H.; Abdulrazak, L.F.; Iqbal, A.; Bashir, M.; Khan, S.; Kim, S. High-performance multiple-input multiple-output antenna system for 5G mobile terminals. *Electronics* **2019**, *8*, 1090. [CrossRef]
2. Abdullah, M.; Kiani, S.H.; Iqbal, A. Eight element multiple-input multiple-output (MIMO) antenna for 5G mobile applications. *IEEE Access* **2019**, *7*, 134488–134495. [CrossRef]
3. Iqbal, A.; Saraereh, O.A.; Ahmad, A.W.; Bashir, S. Mutual coupling reduction using F-shaped stubs in UWB-MIMO antenna. *IEEE Access* **2017**, *6*, 2755–2759. [CrossRef]
4. Acharjee, J.; Mandal, K.; Mandal, S.K. Reduction of mutual coupling and cross-polarization of a MIMO/diversity antenna using a string of H-shaped DGS. *AEU-Int. J. Electron. Commun.* **2018**, *97*, 110–119. [CrossRef]
5. Sindhadevi, M.; Malathi, K.; Henridass, A.; Shrivastav, A.K. Signal integrity performance analysis of mutual coupling reduction techniques using DGS in high speed printed circuit boards. *Wirel. Pers. Commun.* **2017**, *94*, 3233–3249. [CrossRef]
6. Liu, Y.; Yang, X.; Jia, Y.; Guo, Y.J. A low correlation and mutual coupling MIMO antenna. *IEEE Access* **2019**, *7*, 127384–127392. [CrossRef]

7. OuYang, J.; Yang, F.; Wang, Z. Reducing mutual coupling of closely spaced microstrip MIMO antennas for WLAN application. *IEEE Antennas Wirel. Propag. Lett.* **2011**, *10*, 310–313. [CrossRef]

8. Altaf, A.; Alsunaidi, M.A.; Arvas, E. A novel EBG structure to improve isolation in MIMO antenna. In Proceedings of the 2017 USNC-URSI Radio Science Meeting (Joint with AP-S Symposium), San Diego, CA, USA, 9–14 July 2017; pp. 105–106.

9. Yu, A.; Zhang, X. A novel method to improve the performance of microstrip antenna arrays using a dumbbell EBG structure. *IEEE Antennas Wirel. Propag. Lett.* **2003**, *2*, 170–172.

10. Farahani, H.S.; Veysi, M.; Kamyab, M.; Tadjalli, A. Mutual coupling reduction in patch antenna arrays using a UC-EBG superstrate. *IEEE Antennas Wirel. Propag. Lett.* **2010**, *9*, 57–59. [CrossRef]

11. Iqbal, A.; A Saraereh, O.; Bouazizi, A.; Basir, A. Metamaterial-based highly isolated MIMO antenna for portable wireless applications. *Electronics* **2018**, *7*, 267. [CrossRef]

12. Iqbal, A.; Basir, A.; Smida, A.; Mallat, N.K.; Elfergani, I.; Rodriguez, J.; Kim, S. Electromagnetic bandgap backed millimeter-wave MIMO antenna for wearable applications. *IEEE Access* **2019**, *7*, 111135–111144. [CrossRef]

13. Ullah, S.; Yeo, W.H.; Kim, H.; Yoo, H. Development of 60-GHz millimeter wave, electromagnetic bandgap ground planes for multiple-input multiple-output antenna applications. *Sci. Rep.* **2020**, *10*, 1–12. [CrossRef] [PubMed]

14. Habashi, A.; Nourinia, J.; Ghobadi, C. Mutual coupling reduction between very closely spaced patch antennas using low-profile folded split-ring resonators (FSRRs). *IEEE Antennas Wirel. Propag. Lett.* **2011**, *10*, 862–865. [CrossRef]

15. Lee, J.Y.; Kim, S.H.; Jang, J.H. Reduction of mutual coupling in planar multiple antenna by using 1-D EBG and SRR structures. *IEEE Trans. Antennas Propag.* **2015**, *63*, 4194–4198. [CrossRef]

16. Hafezifard, R.; Naser-Moghadasi, M.; Mohassel, J.R.; Sadeghzadeh, R. Mutual coupling reduction for two closely spaced meander line antennas using metamaterial substrate. *IEEE Antennas Wirel. Propag. Lett.* **2015**, *15*, 40–43.

17. Ghosh, J.; Ghosal, S.; Mitra, D.; Bhadra Chaudhuri, S.R. Mutual coupling reduction between closely placed microstrip patch antenna using meander line resonator. *Prog. Electromagn. Res.* **2016**, *59*, 115–122. [CrossRef]

18. Babu, K.V.; Anuradha, B. Design of Wang shape neutralization line antenna to reduce the mutual coupling in MIMO antennas. *Analog Integr. Circuits Signal Process.* **2019**, *101*, 67–76. [CrossRef]

19. Vishvaksenan, K.S.; Mithra, K.; Kalaiarasan, R.; Raj, K.S. Mutual coupling reduction in microstrip patch antenna arrays using parallel coupled-line resonators. *IEEE Antennas Wirel. Propag. Lett.* **2017**, *16*, 2146–2149. [CrossRef]

20. Luan, H.; Chen, C.; Chen, W.; Zhou, L.; Zhang, H.; Zhang, Z. Mutual Coupling Reduction of Closely E/H-Plane Coupled Antennas Through Metasurfaces. *IEEE Antennas Wirel. Propag. Lett.* **2019**, *18*, 1996–2000. [CrossRef]

21. Khalid, M.; Iffat Naqvi, S.; Hussain, N.; Rahman, M.; Fawad; Mirjavadi, S.S.; Khan, M.J.; Amin, Y. 4-Port MIMO antenna with defected ground structure for 5G millimeter wave applications. *Electronics* **2020**, *9*, 71. [CrossRef]

22. Sehrai, D.A.; Abdullah, M.; Altaf, A.; Kiani, S.H.; Muhammad, F.; Tufail, M.; Irfan, M.; Glowacz, A.; Rahman, S. A Novel High Gain Wideband MIMO Antenna for 5G Millimeter Wave Applications. *Electronics* **2020**, *9*, 1031. [CrossRef]

23. Iqbal, A.; Smida, A.; Alazemi, A.J.; Waly, M.I.; Mallat, N.K.; Kim, S. Wideband Circularly Polarized MIMO Antenna for High Data Wearable Biotelemetric Devices. *IEEE Access* **2020**, *8*, 17935–17944. [CrossRef]

24. Kumar, P.; Urooj, S.; Alrowais, F. Design of quad-port MIMO/Diversity antenna with triple-band elimination characteristics for super-wideband applications. *Sensors* **2020**, *20*, 624. [CrossRef] [PubMed]

25. Mohamadzade, B.; Lalbakhsh, A.; Simorangkir, R.B.; Rezaee, A.; Hashmi, R.M. Mutual Coupling Reduction in Microstrip Array Antenna by Employing Cut Side Patches and EBG Structures. *Prog. Electromagn. Res.* **2020**, *89*, 179–187. [CrossRef]

26. Margaret, D.H.; Subasree, M.; Susithra, S.; Keerthika, S.; Manimegalai, B. Mutual coupling reduction in MIMO antenna system using EBG structures. In Proceedings of the 2012 International Conference on Signal Processing and Communications (SPCOM), Bangalore, India, 22–25 July 2012; pp. 1–5.

27. Si, L.; Jiang, H.; Lv, X.; Ding, J. Broadband extremely close-spaced 5G MIMO antenna with mutual coupling reduction using metamaterial-inspired superstrate. *Opt. Express* **2019**, *27*, 3472–3482. [CrossRef] [PubMed]

28. Wei, K.; Li, J.Y.; Wang, L.; Xing, Z.J.; Xu, R. Mutual coupling reduction by novel fractal defected ground structure bandgap filter. *IEEE Trans. Antennas Propag.* **2016**, *64*, 4328–4335. [CrossRef]

29. Kiani, S.H.; Mahmood, K.; Altaf, A.; Cole, A.J. Mutual coupling reduction of MIMO antenna for satellite services and radio altimeter applications. *Int. J. Adv. Comput. Sci. Appl.* **2018**, *9*, 23–26. [CrossRef]

30. Arun, H.; Sarma, A.K.; Kanagasabai, M.; Velan, S.; Raviteja, C.; Alsath, M.G.N. Deployment of modified serpentine structure for mutual coupling reduction in MIMO antennas. *IEEE Antennas Wirel. Propag. Lett.* **2014**, *13*, 277–280. [CrossRef]

31. Balanis, C.A. *Antenna Theory: Analysis and Design*; John Wiley & Sons: Hoboken, NJ, USA, 2016.

32. Iqbal, A.; Selmi, M.A.; Abdulrazak, L.F.; Saraereh, O.A.; Mallat, N.K.; Smida, A. A Compact Substrate Integrated Waveguide Cavity-Backed Self-Triplexing Antenna. *IEEE Trans. Circuits Syst. II Express Briefs* **2020**. [CrossRef]

33. Elfergani, I.; Iqbal, A.; Zebiri, C.; Basir, A.; Rodriguez, J.; Sajedin, M.; Pereira, A.D.O.; Mshwat, W.; Abd-Alhameed, R.; Ullah, S. Low-Profile and Closely Spaced Four-Element MIMO Antenna for Wireless Body Area Networks. *Electronics* **2020**, *9*, 258. [CrossRef]

34. Farahani, M.; Pourahmadazar, J.; Akbari, M.; Nedil, M.; Sebak, A.R.; Denidni, T.A. Mutual coupling reduction in millimeter-wave MIMO antenna array using a metamaterial polarization-rotator wall. *IEEE Antennas Wirel. Propag. Lett.* **2017**, *16*, 2324–2327. [CrossRef]

35. Qamar, Z.; Naeem, U.; Khan, S.A.; Chongcheawchamnan, M.; Shafique, M.F. Mutual coupling reduction for high-performance densely packed patch antenna arrays on finite substrate. *IEEE Trans. Antennas Propag.* **2016**, *64*, 1653–1660. [CrossRef]

Article

A Novel Asymmetric Patch Reflectarray Antenna with Ground Ring Slots for 5G Communication Systems

M. Hashim Dahri [1], **M. Haizal Jamaluddin** [2,*], **Fauziahanim C. Seman** [1,*],
Muhammad Inam Abbasi [3], **Adel Y. I. Ashyap** [1], **M. Ramlee Kamarudin** [1] **and Omar Hayat** [4]

[1] Faculty of Electrical and Electronic Engineering, Universiti Tun Hussein Onn Malaysia (UTHM),
    Batu Pahat 86400, Malaysia; muhammadhashimdahri@yahoo.com (M.H.D.);
    ashyap2007@gmail.com (A.Y.I.A.); mramlee@uthm.edu.my (M.R.K.)
[2] Wireless Communication Centre of Universiti Teknologi Malaysia (UTM), Johor Bahru 81310, Malaysia
[3] Faculty of Electrical and Electronic Engineering Technology, Universiti Teknikal Malaysia Melaka (UTeM),
    Melaka 76100, Malaysia; inamabbasi@utem.edu.my
[4] Department of Engineering, National University of Modern Languages (NUML), Islamabad 44000, Pakistan;
    ohayat@numl.edu.pk
*   Correspondence: haizal@fke.utm.my (M.H.J.); fauziahs@uthm.edu.my (F.C.S.)

Received: 25 July 2020; Accepted: 29 August 2020; Published: 5 September 2020

Abstract: The narrow bandwidth and low gain performances of a reflectarray are generally improved at the cost of high design complexity, which is not a good sign for high-frequency operation. A dual resonance asymmetric patch reflectarray antenna with a single layer is proposed in this work for 5G communication at 26 GHz. The asymmetric patch is developed from a square patch by tilting its one vertical side by a carefully optimized inclination angle. A progressive phase range of 650° is acquired by embedding a circular ring slot in the ground plane of the proposed element for gain improvement. A 332-element, center feed reflectarray is designed and tested, where its high cross polarization is suppressed by mirroring the orientation of asymmetric patches on its surface. The asymmetric patch reflectarray offers a 3 dB gain bandwidth of 3 GHz, which is 4.6% wider than the square patch reflectarray. A maximum measured gain of 24.4 dB has been achieved with an additional feature of dual linear polarization. Simple design with wide bandwidth and high-gain of asymmetric patch reflectarray make it suitable to be used in 5G communications at high frequencies.

Keywords: 5G; asymmetric patch; dual polarization; dual resonance; reflectarray; unit cell

---

1. Introduction

Antennas for 5G can be divided into two categories, one that is required at the user end and a second that is needed at the base station. In both of these categories, Multiple Input Multiple Output (MIMO) antennas have been seen as potential candidates to fill the requirement [1,2]. MIMOs are actually the combination of many same types of antennas to produce a required beam pattern, which is not easy to get with a single antenna. This combination of many antennas increases the design complexity and cost of the system. The user end, due to its limited space, is a suitable place to be designed with a MIMO antenna. The base station, on the other hand, can be seen as a potential place where different types of antennas can be tested for a required 5G performance. The selection of an antenna at the base station mainly depends on the range of its acquired parameters that are needed for 5G communication [3]. Apart from a selected operating frequency range, a gain of more than 20 dB and a bandwidth of more than 1 GHz are required for a 5G antenna that is supposed to be used as a transmitter at the base station [4,5]. The additional and optional features of a 5G antenna also include polarization diversity and adaptive beamsteering [6]. Polarization diversity could be handy to overcome the flaw of narrow bandwidth, while swift electronic beamsteering can reduce the number of antennas on the base station. The antenna operating frequency for 5G operation is mainly considered in the range of either millimeter wave or sub-millimeter wave [7]. The common millimeter wave frequency ranges that are used for 5G-based research are mostly above 22 GHz. Array antennas, due to their ability of acquiring high-gain, have been widely proposed by the researchers to be used in the millimeter wave frequency range for 5G communication [8]. Phased arrays and reflectarrays both can be good candidates for a 5G base station antenna. Between them, reflectarray does not require a phase shifter circuit, which significantly reduces its cost and design complexity as compared to a phased array antenna.

A microstrip reflectarray is generally considered as a planar phased array with performance features of the parabolic reflector [9]. It was introduced to overcome the flaws of the curvy design of the parabolic reflector and the high complexity of the phased array [10]. The collimation of incident signals from the surface of the reflectarray is performed by properly designing its unit cell element. Most of the recent works proposing 5G reflectarrays have worked on the enhancement of its bandwidth performance. However, in reality, acquiring a bandwidth performance of more than 1 GHz at millimeter wave range with reflectarray antenna is not a big issue. Its gain can also be increased by using a large aperture size. The main issue with a reflectarray antenna that limits its progression for 5G communication is its design complexity. The enhanced performance of a reflectarray antenna and its design complexity are complementary to one another. A simple design would also reduce its cost and make it suitable to be produced in large quantities.

Reflectarray performance can be enhanced by considering the modification in the design of its unit cell element [3]. A unit cell is comprised of a conducting patch element printed on a grounded dielectric substrate. The substrate and patch both can be considered for possible design adjustments for performance improvement [11]. A wide reflection phase (S_{11} phase) range of unit cell element ensures a good bandwidth performance. The substrate of the unit cells can be used as stacked layers, or an air gap can be inserted in it before the ground plane to cope with the narrow bandwidth issue [12–14]. However, this could increase the design efforts and make it difficult to be fabricated, especially at millimeter wave frequencies. On the other hand, different patch element designs, such as a combination of elements [15], fractal elements [16], and multiple phase tuning stubs [17] can also be used for the performance enhancement. Among them, the last two designs are only suitable for low frequency operation due to their design complexity, and the first one can face mutual coupling issues due to more than one element being used in it [3].

The primary purpose of this research work is to produce a reflectarray unit cell element with dual resonance response without using a dual patch or dual layer structure. This dual resonance response is further utilized for an extended reflection phase range to acquire a wider bandwidth performance. Dual resonance response with a single layer can also be acquired with a Fractal type element. However, Fractal elements have a very complicated design structure; that is why it is challenging to implement them at very short wavelengths of high frequencies [3]. Additionally, it is normal to use an air gap in the substrate of a reflectarray antenna for bandwidth improvement, which increases the design efforts. The addition of this air gap with a minor fabrication error can cause a huge deflection in the resonance of the reflectarray antenna at high frequencies. Therefore, this type of air gap is not used in this design of asymmetric patch reflectarray antenna, which relies totally on a new shape of the patch element to generated dual resonance response.

A similar work published in [4] also proposes a reflectarray antenna for 5G communications. The work reported in [4] contains reflectarray of conventional circular ring elements with full ground, whereas this work offers a novel asymmetric patch reflectarray element with ground ring slots for bandwidth and gain enhancement for 5G application. Moreover, this work also offers an improved reflectarray operation especially in terms of a smaller size, wider reflection phase range and additional polarization as compared to the work published in [4]. The detailed comparison between this work and [4] has been presented alongside with many other related works in Section 5 of this paper.

The design complexity is always a major factor if a reflectarray antenna is proposed for 5G operation. Therefore, a simple design of the unit cell patch element is required with a wide reflection phase range and suitable aperture size for reflectarray compatibility with 5G communication. In this work, a single layer asymmetric patch unit cell element is first time introduced to enhance the reflectarray performance at 26 GHz frequency. The main novelty of the proposed work is the introduction of a new asymmetric patch reflectarray element that can produce two close resonances for reflection phase range enhancement without using an extra resonant layer or resonant element. A simple design of the proposed asymmetric patch element that can be easily fabricated at short wavelengths of high frequencies is also considered as a huge advantage. The asymmetry in the design of the patch element is used as a tactic to enhance its reflection phase range. Asymmetry also creates a high level of unwanted cross polarization [18], which is tackled by selecting a proper element orientation on the reflectarray surface. Simulations of unit cell element and full reflectarray are performed by CST Microwave Studio within a frequency range of 24 to 28 GHz. The selected frequency range is considered due to its application in 5G communications [4,19]. The proposed design of the reflectarray antenna has also been experimentally verified. In the end, the proposed reflectarray antenna with its acquired parameters has been used as an indoor base station to measure its signal strength at various locations for 5G communication.

2. Asymmetric Patch Unit Cell Element

The asymmetric patch element has been developed from a square patch element that is taken as the primary structure to construct it. A low loss Rogers Rt/D 5880 material with a dielectric constant of 2.2 and a thickness of 0.254 mm is selected to hold the grounded asymmetric patch element. The substrate length and width have been set to be $\lambda/2$, which is 5.77 mm at 26 GHz resonant frequency. The evolution of an asymmetric patch element from a standard square patch is demonstrated in Figure 1a. It can be seen from Figure 1a that a vertical side of the square patch element is tilted to some optimized inclination angle (θ). Due to the tilting of a side of the patch element, two different lengths (L_1 and L_2) have been introduced in the same patch element. The resonance of a microstrip patch elements depends on the dimension of its length. Therefore, in the case of an asymmetric patch element, two different resonances can be generated from a single element as given in the following equation [20].

$$f_1 = \frac{c}{2L_1 \sqrt{\varepsilon_{eff}}} \; and \; f_2 = \frac{c}{2L_2 \sqrt{\varepsilon_{eff}}} \tag{1}$$

Here, c is the speed of light, ε_{eff} is the effective dielectric constant of the substrate, while f_1 and f_2 are the two resonant frequencies associated with the two different lengths L_1 and L_2 of the asymmetric patch element.

A standard procedure for the simulation of a reflectarray unit cell element is adopted that is based on the large array approach, as explained in [21]. In the simulation method, the reflectarray unit cell is enclosed by the periodic boundary conditions to generate a large array approach. The periodic boundaries are comprised of pure magnetic and pure electric walls on right-left and top-bottom sides of the unit cell element respectively. A waveguide port is placed in front of the unit cell element at a distance of quarter wavelength to observe the reflective characteristics of the unit cell element. The boundary at the ground plane of the element is kept open as it backs the free space behind the reflectarray unit cell element. The dual resonance phenomenon of the asymmetric patch element can be seen from Figure 2a, where simulated reflection loss (S_{11} magnitude) of the asymmetric patch element has been plotted with respect to the inclination angle (θ). It can be observed that when the inclination angle is decreased beyond 86°, an extra second resonance has been formed in the frequency response of the asymmetric patch element. The two resonances shown in Figure 2 are occurred due to the asymmetric design of the unit cell patch element. At this optimization stage the center frequency of the first resonance is 25.5 GHz. Both the resonances offer their respective reflection phases of more than 300° and a collective reflection phase span of more than 600°. The asymmetric patch element without an inclination angle ($\theta = 0°$) would be a square patch element with an ordinary frequency response of a single resonance.

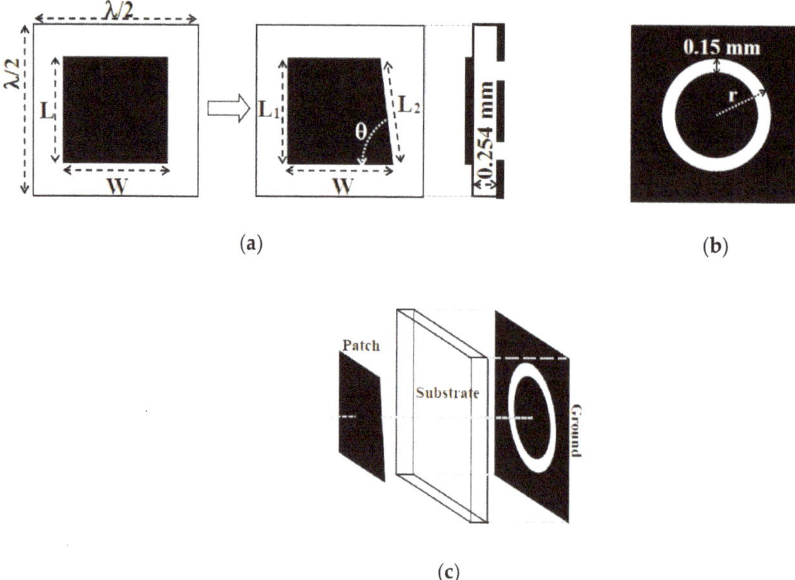

Figure 1. (**a**) Development of asymmetric patch element from a square patch element; (**b**) Ground ring slot; (**c**) Alignment of the asymmetric patch with ground ring slot.

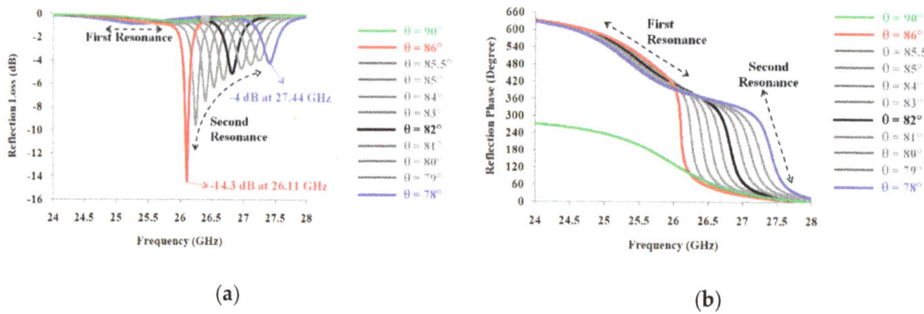

Figure 2. Simulated frequency response of the asymmetric patch element with variable inclination angle and full ground plane: (**a**) Reflection loss; (**b**) Reflection phase.

The dual resonance response, as driven in Figure 2, still lacks the linearity in the reflection phase that is an important factor in getting an enhanced bandwidth performance. In order to get this linearity in the reflection phase, optimization has been performed with the variable length ($L = L_1$) and variable width (W) of the asymmetric patch element. The simulated results of this optimization have been shown in Figure 3. It can be analyzed from Figure 3a,b that the slope of the reflection phase is a function of change in length and width of the asymmetric patch element, respectively. Additionally, a linear reflection phase curve can be obtained with the asymmetric patch element if its dimensions are selected properly. These dimensions can be varied a little further to achieve the same resonant frequency when a ring slot is embedded in the ground plane. The optimized parameters of the asymmetric patch element at 26 GHz with the full ground and with circular ring slot in the ground are summarized in

Table 1. Values in the second column of Table 1 belong to the optimization performed in Figure 3, and values in the third column of Table 1 are obtained by varying values of the second column when a ring slot is added in the ground plane.

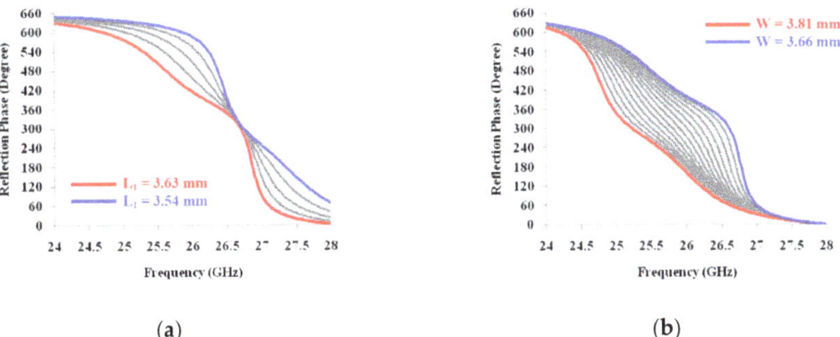

(a)   (b)

Figure 3. Optimization of the slope of the reflection phase curve of the asymmetric patch element with respect to: (**a**) Variable length (L_1); (**b**) Variable width (W).

Table 1. Optimized dimensions of the asymmetric patch element at 26 GHz.

Parameter	With Full Ground	With Ground Ring Slot (r = 1 mm)
L_1 (mm)	3.59	3.15
L_2 (mm)	3.61	3.18
W (mm)	3.77	3.33
θ (°)	83.5	82.6

The progressive phase distribution in the reflectarray can normally be acquired with the variable size of its patch elements or by applying the element rotation technique [22,23]. However, it would be difficult to change the overall size of the asymmetric patch element or rotate it while keeping its shape intact. Therefore, a circular ring slot, as depicted in Figure 1b, has been embedded in the ground plane of the asymmetric patch element. The variable radius (r) of this grounded embedded ring slot has been used to progressively vary the reflection phase response of the asymmetric patch element [24]. Additionally, the ground ring slot can easily support the dual linear polarization response of the symmetric patch reflectarray that is an essential parameter for 5G communications. The same dual linear polarization could also have been obtained by a square ring slot, but the square ring slot attains a larger defecting area than a circular ring slot. A larger defecting area in ground plane would result in an increase in the reflection loss performance, which is not good for a high-gain reflectarray operation. The width of this ground ring slot should be kept as minimum as possible in order to suppress the generation of the back radiations [24,25]. As depicted in Figure 1b, the width of the ring slot is kept constant at 0.15 mm, because it is the minimum possible dimension that can be fabricated using available laboratory resources. Figure 1c shows the alignment of the asymmetric patch element with the ground ring slot. It can be seen from Figure 1c that the patch is situated in the center of the substrate that aligns directly with the center of the ring slot. All the control parameters that affect the results of asymmetric patch element have been mentioned in Figures 1–3 and Table 1. The remaining parameters have been kept constant throughout the simulation process as they are not used to perform any change in the response of the asymmetric patch element. Those constant parameters include boundary conditions, waveguide port distance, substrate properties and substrate dimensions.

2.1. Asymmetry and Cross Polarization

The asymmetric patch element has an inclined side instead of a vertical one. In this scenario, a vertically polarized incident electric field (E_i) generates a diagonally flowing surface current (J), as shown in Figure 4a. The diagonal movement of surface current breaks it up into a vertical (J_y) and a horizontal (J_x) component. These surface current components induce a reflected electric field (E_{ry} and E_{rx}) in the same manner. The reflected electric field component $\vec{E}_{ry}$ is called as co-polarized reflected electric field because it is in the same polarization with the incident electric field. Alternatively, E_{rx} is called as cross-polarized reflected electric field because it is in the orthogonal polarization with the incident electric field.

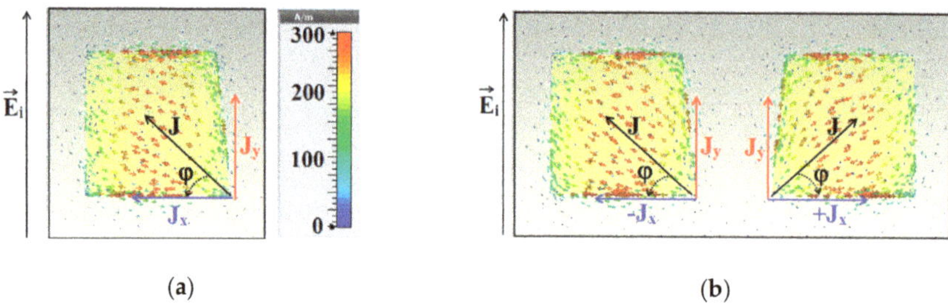

(a)   (b)

Figure 4. (**a**) Surface current flow on asymmetric patch element; (**b**) Two asymmetric patch elements in mirror orientation.

The effect of high cross polarization can be eliminated by mirroring the orientation of the elements on the reflectarray surface [26]. The depiction of two mirror elements, which are horizontally flipped, is shown in Figure 4b. As illustrated in Figure 4b, the two horizontal components of surface current ($+J_x$ and $-J_x$) are in the opposite direction to one another. The positive and negative signs of $\vec{J}_x$ are assigned to represent their opposite directions. In this scenario, the reflected electric field for two asymmetric patch elements can be calculated as described in the following derivation.

$$\vec{E}_r + \vec{E}_r = \vec{E}_{rx} + \vec{E}_{ry} - \vec{E}_{rx} + \vec{E}_{ry}$$
$$2\vec{E}_r = 2\vec{E}_{ry}$$
$$\vec{E}_r = \vec{E}_{ry} = E_r \sin\varphi \tag{2}$$

Equation (2) shows that the cross-polarized component of the reflected electric field can be eliminated if the asymmetric patch elements are arranged in mirror orientation to each other on the surface of reflectarray.

2.2. Experimental Results of Asymmetric Patch Element

The experimental verification of the asymmetric patch unit cell element with embedded ground ring slots has been performed using scattering parameter measurements. The fabricated unit cell elements that have been used for the experimental purpose are shown in Figure 5a. There are total of 10 different fabricated samples of the unit cells that have been used for the measurements. Change in the ground ring radius for the fabricated unit cells can be spotted from Figure 5a, while all the asymmetric patch elements have the same size. The unit cell element contains two asymmetric patches, which makes it suitable to be aligned with the open end of the waveguide simulator. The waveguide simulator is a piece of the standard equipment used for the scattering parameter measurements of a reflectarray unit cell element [27]. The rectangular dimensions of the waveguide simulator are obtained by considering it for the wave propagation of TE$_{10}$ mode. The unit cell element cannot be displaced or moved during the measurement as it is fixed within the open face cavity of the waveguide. Therefore, no deflection in the measured results is supposed to happen due to the mounting or placement of the unit cell element. The unit cell measurement setup is depicted in Figure 5b, where the waveguide simulator is connected with a Vector Network Analyzer (VNA) through a standard WR-34 coaxial to waveguide adapter. A standard process of the VNA calibration using open, short and load connectors has been performed to properly calibrate the measurement system. The predefined dimensions of the waveguide simulator and WR-34 adapter make them suitable to be used from 22 to 33 GHz frequencies. Although, the scattering parameter measurements have been performed from 24 to 28 GHz.

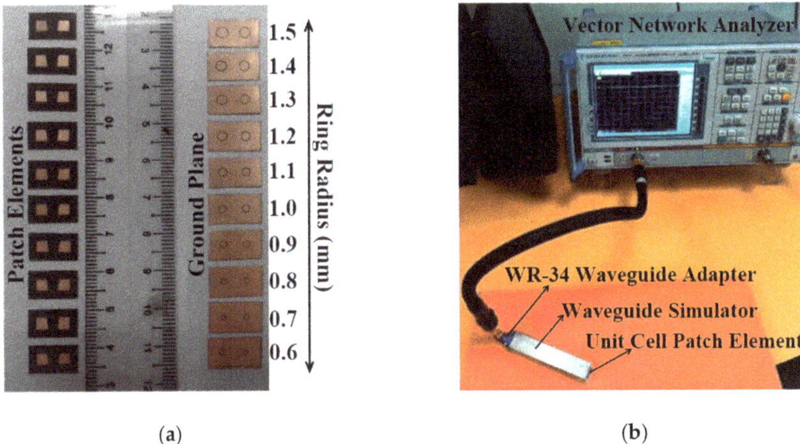

Figure 5. (**a**) Fabricated unit cells of asymmetric patch element with a variable radius of ground ring slots; (**b**) Unit cell measurement setup.

Figure 6a,b show the graphs of the reflection response of the asymmetric patch element with the full ground and with ground ring slot, respectively. A dual resonance response can be spotted in both the mentioned cases. Asymmetric patch with full ground is observed to offer measured resonances at 25.5 GHz and 26.25 GHz, whereas its counterpart with embedded ground ring slot attains measured resonances at 25.5 GHz and 26.75 GHz. The asymmetric patch element with and without ground ring slot has the dimensions that have been mentioned in Table 1. However, as expected, a higher reflection loss performance has been observed from the asymmetric patch element with the ground ring slot as compared to a full ground element. The leakage currents in the ground ring slot are the main reason behind this higher reflection loss. Additionally, a higher measured loss as compared to a simulated one and a small deflection in the measured resonant frequencies are observed that occurs because of the unavoidable effects of the fabrication errors, waveguide, connectors and cables used during the measurements process. The full ground and slotted ground asymmetric patch elements attain a measured linear reflection phase range of 480° and 510°, respectively. This wide reflection phase range is essential for wideband performance in reflectarray antenna.

The progressive phase distribution of the asymmetric patch element at 26 GHz with the full ground and with the slotted ground is depicted in Figure 7a,b, respectively. It can be analyzed that, the asymmetric patch element with the full ground does not acquire a wideband performance for the frequencies other than 26 GHz. It is because of the sensitive nature of the asymmetric patch element, as a full phase span requires just a change of 0.2 mm in length (L_1), which is practically hard to differentiate. Alternatively, a ground ring slot serves the same purpose with a change in radius of 0.9 mm, which is less sensitive than its counterpart. Figure 7c shows the reflection phase response of the asymmetric patch element with the ground ring slot for different incident angles. A full phase swing can be observed up to 32° of incident angle, which is the maximum value in the proposed design of the reflectarray antenna of this work.

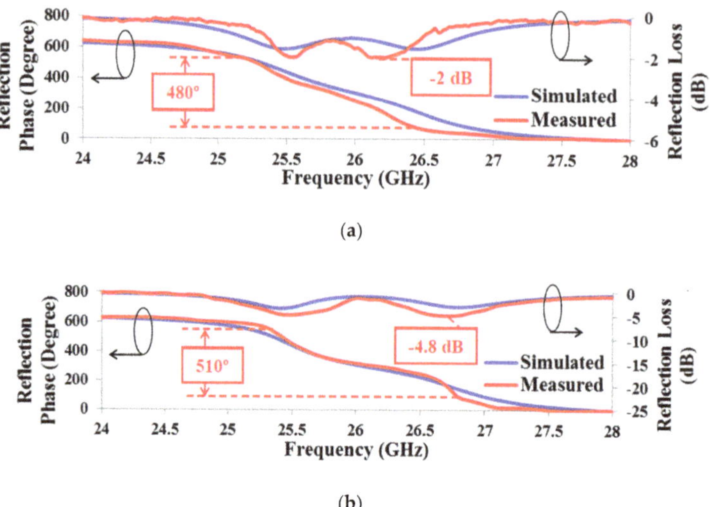

Figure 6. Simulated and measured reflection response of asymmetric patch element with: (**a**) Full ground; (**b**) Ground ring slot with r = 1 mm.

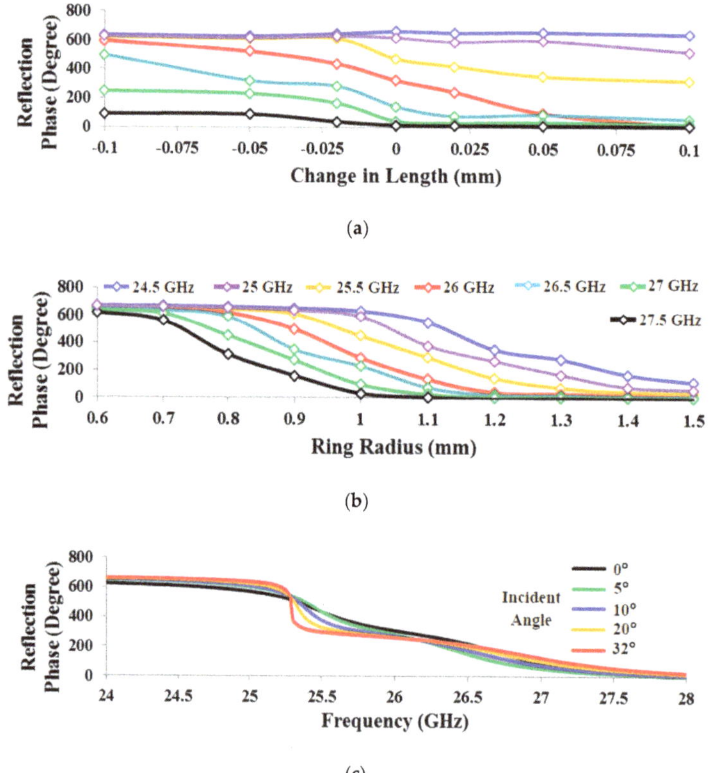

Figure 7. Reflection phase response of asymmetric patch element: (**a**) Measured with full ground; (**b**) Measured with ground ring slot; (**c**) Simulated with the different incident angle at 26 GHz.

3. Reflectarray of Asymmetric Patch Elements

A standard procedure of designing the microstrip reflectarray antenna with its proper experimental validation is explained in [10,22,28,29]. The same process has been followed to design and validate the asymmetric patch reflectarray antenna with a variable radius of its ground ring slots. The elements of a reflectarray are needed to be properly designed with an accurate progressive phase distribution to get proper collimation of its reflected signals. The progressive phase distribution on the surface of a reflectarray antenna is calculated using Equation (3) [10].

$$\Delta\varphi = \varphi - \varphi_i = \frac{2\pi}{\lambda}(f - f_i) \tag{3}$$

where $\Delta\varphi$ is the change in reflection phase, φ is the progressive phase center, φ_i is the reflection phase of a selected element, f is the focal length of the feed and f_i is the distance of the selected element from the center of the feed. This mathematical relation has been used to design a 332 element asymmetric patch reflectarray antenna with a variable radius of its ground ring slots that are responsible for attaining an accurate progressive phase distribution. The distance between the asymmetric patches is $\lambda/2$, which is uniform for all the elements on the reflectarray. The variability in the ring slot radius is responsible for making the design periodicity in reference to the feed position to properly eliminate the effect of variable path delays that occur due to a flat surface of the reflectarray. The feed power is selected as 10 dB, which offers a maximum aperture illumination at an f/D of 0.8 [11]. A common wet etching fabrication process with high precision is used to make the proposed designs of the asymmetric patch reflectarray antenna. The fabricated designs with different element orientations that are shown in Figure 8 have been used for experimental verification. Figure 8a,b depict the two fabricated designs of the asymmetric patch reflectarray without and with mirror orientation of the elements, respectively. Figure 8b also depicts the progression of the reflection phase on the surface of the reflectarray that is acquired by a variable ring radius for proper collimation of the reflected signals. The ring radius values are also depicted alongside its respective phase values. The size of all the asymmetric patch elements has been kept constant in both of the designs, which is previously mentioned in Table 1.

The experimental results in comparison with the simulated ones are plotted in Figure 9. The radiation patterns with the measured graphs of cross polarization of non-mirror and mirror orientation of the asymmetric patch elements have been shown in Figure 9a,b, respectively. An improvement in the cross polarization level has been noticed from −5 dB to −20 dB when the asymmetric patch elements are mirrored on the surface of the reflectarray. The cross polarization level of −20 dB or lower can be referred to as good isolation between the co-polarization and cross polarization of an antenna [30]. Simulated cross polarization graphs are ignored here for graph clarity; however, their values are provided. The simulated and measured results are easily comparable with minor discrepancies.

Figure 9c contains measured E-plane graphs for mirror element orientation at different frequencies to show the reliability of the radiation patterns. These results of the asymmetric patch reflectarray antenna have been taken using a vertically polarized incident electric field. Gain versus frequency curves of asymmetric patch reflectarray antenna with vertical polarization are plotted in Figure 10. It can be seen that the reflectarray without mirror orientation of elements offers higher gain than the reflectarray with mirror orientation of elements. This is due to a higher magnitude of the tilted reflected electric field as compared to a lower magnitude of the vertical electric field, as explained in Section 2.1. However, this high-gain performance is idle due to a high cross polarization when elements are not mirrored to each other. A significant improvement in gain is also noticed when a full ground plane is replaced with the circular ring slots. In this case, the gain has been improved from 18.8 dB to 24.4 dB, with increment in aperture efficiency from 7.7% to 28%. The measured 3 dB gain bandwidth of asymmetric patch reflectarray with mirror element orientation is 3 GHz, which covers 75% of the selected frequency band of operation.

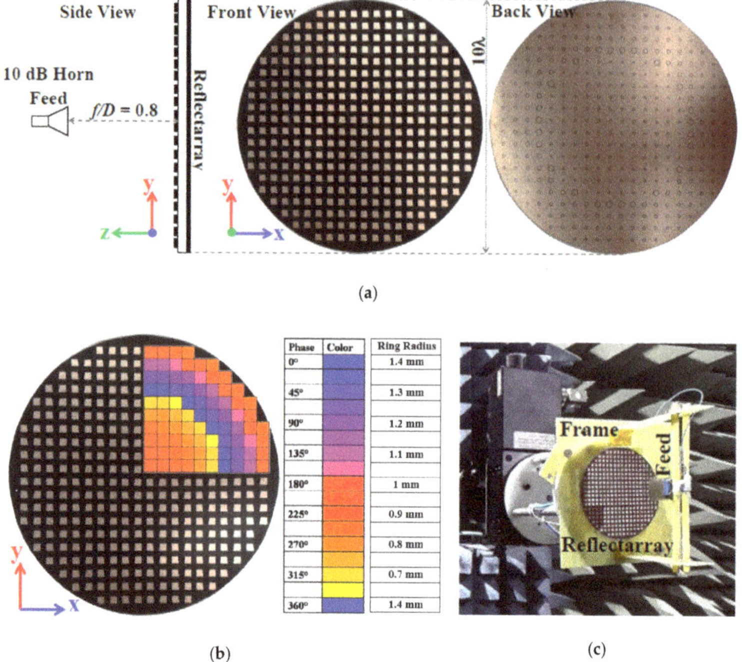

(a)

(b)

(c)

Figure 8. Asymmetric patch reflectarray antenna: (**a**) Non mirror orientation of elements and ground ring slots; (**b**) Mirror orientation of elements; (**c**) During measurements.

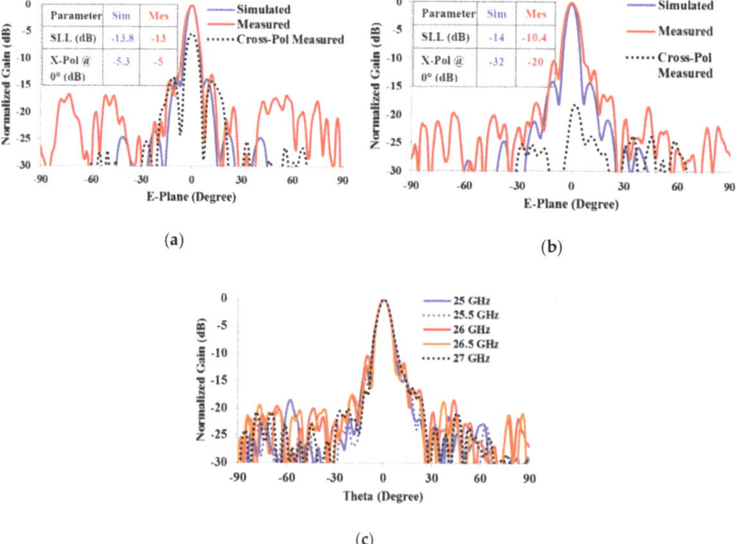

(a)

(b)

(c)

Figure 9. Vertically polarized radiation pattern results of asymmetric patch reflectarray with embedded ground slots for: (**a**) Non mirror elements at 26 GHz; (**b**) Mirror elements at 26 GHz; (**c**) Measured E-plane with mirror elements at different frequencies.

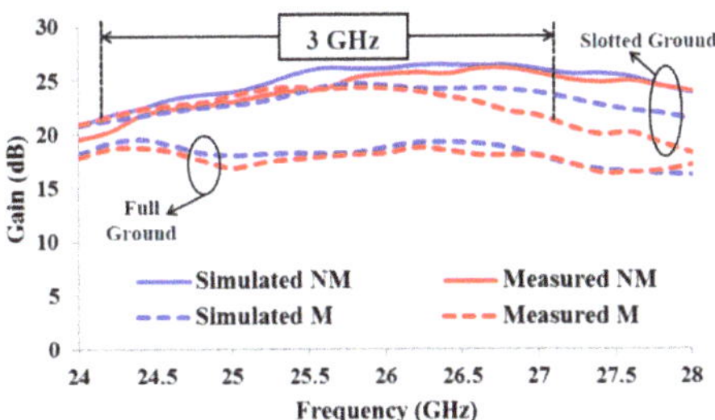

Figure 10. Gain versus frequency response for vertical polarization of asymmetric patch element reflectarray with and without embedded ground slots (NM = Non-Mirror and M = Mirror element orientation).

In order to analyze the polarization diversity performance of the asymmetric patch reflectarray antenna of ground ring slots, it is also measured with a horizontally polarized incident electric field. Simulated and measured radiation pattern graphs with mirror orientation of asymmetric patch elements have been plotted in Figure 11a. The reduction in cross polarization due to mirror orientation of asymmetric patch elements has also been observed in Figure 11a. Figure 11b shows gain with respect to the frequency of the asymmetric patch reflectarray antenna with horizontal polarization. A minor reduction in the maximum gain performance of the reflectarray antenna has been observed due to change in its polarization operation. The measured gain with horizontal polarization is 23.9 dB, which is 0.5 dB less than its orthogonal polarization counterpart and offers an aperture efficiency of 25%. Alternatively, the 3 dB gain bandwidth is increased up to 3.6 GHz that is 13.8% at 26 GHz. Moreover, an improvement in the measured side lobe level has also been noticed, as depicted in Figure 11a. The attained results of the asymmetric patch reflectarray antenna with horizontal polarization also falls well within the range of required parameters needed for 5G communication. It can be said that the proposed reflectarray antenna supports the polarization diversity in its operation and offers good performance with both vertical and horizontal polarization.

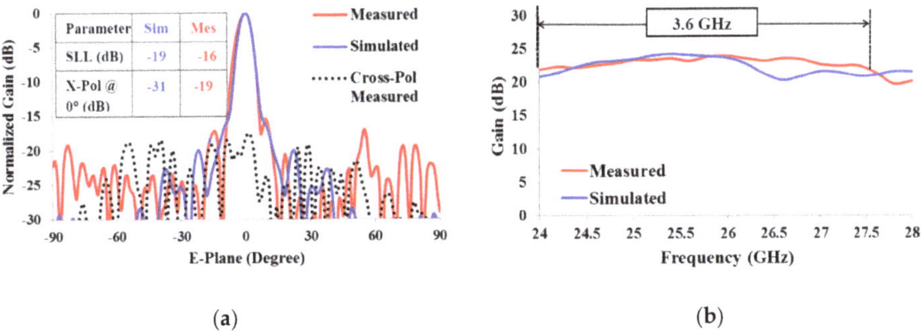

(a)   (b)

Figure 11. Horizontally polarized results of asymmetric patch reflectarray with embedded ground slots for mirror elements (**a**) Radiation pattern at 26 GHz (**b**) Gain versus frequency response.

4. Receive Signal Strength (RSS) Measurement

In 5G communications, the received signal strength (RSS) and received signal quality (RSQ) are measured to determine the probability of cell selection and handover [31]. A high value of RSS would mean to have a high power cell with low interference from neighboring cells and low noise level. Therefore, it can also increase the chances of high quality signal reception at the user equipment. A detailed analysis of RSS measurement for 5G base stations with signal coverage improvement is provided in [32], where two high power horn antennas are used as the transmitter (Tx) and receiver Rx. It is believed that a high value of RSS can provide a high RSQ that is essential for better signal reception in an indoor microcell environment. Therefore, the use of a wideband and high-gain antenna, such as reflectarray, at 5G base station would definitely improve the quality of signal detection at the user equipment. Therefore, the proposed asymmetric patch reflectarray antenna is measured here as an indoor 5G base station and its results are compared with a conventional horn antenna to a draw a fair comparison between them.

The proposed asymmetric patch reflectarray antenna is tested as an indoor base station to measure the receive signal strength at a variable distance. The overall setup for the measurement is explained in [31], where a horn antenna was used as a transmitter at the base station, which is replaced here with the reflectarray antenna. Table 2 summarizes the characteristics and values of the measurement setup. An operating frequency of 26 GHz is selected to measure the dual linear polarized response of the reflectarray antenna. A standard 20 dB horn antenna (CB-28-20-C-KF) is used as a receiver (Rx). The distance between transmitter (Tx) and receiver is varied from 1 m to 16 m with a constant height of 2 m. The minimum distance of 1 m is greater than the minimum far-field distance, which is 2λ, of the Tx antenna. The maximum distance of 16 m is the longest indoor corridor in the Wireless Communication Centre building. Figure 12 shows the Tx and Rx antennas during the measurements.

Table 2. Characteristics of RSS measurement setup.

Parameter	Value or Characteristic
Tx Power	23 dBm
Carrier Frequency	26 GHz
Tx Antenna	Reflectarray (This Work)
Rx Antenna	20 dB Standard Horn
Tx/Rx Height	2 m
Tx Polarization	Dual Linear
Distance	1 m to 16 m
Tx Antenna Beamwidth	6°
Rx Antenna Beamwidth	14°

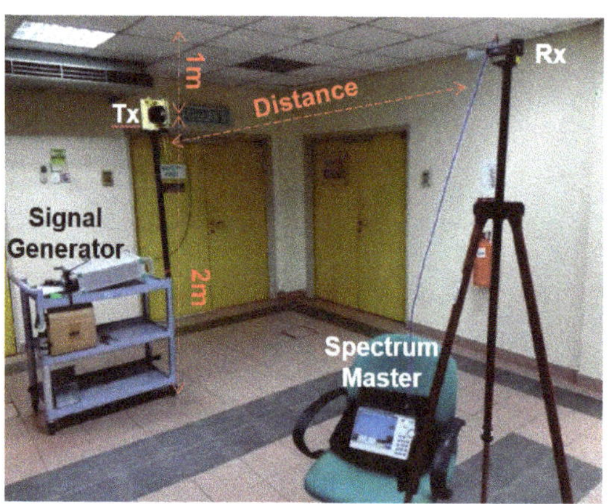

Figure 12. Indoor RSS measurement setup for asymmetric patch reflectarray antenna.

A signal generator (Anritsu MG3694C) is used before the Tx antenna and a spectrum master (Anritsu MS2720T) is connected to the Rx antenna for signal generation and detection, respectively. The main purpose of this study is to analyze the improvement in RSS for an urban indoor microcell sector with a reflectarray antenna. Measured RSS of the asymmetric patch reflectarray antenna and a standard horn antenna as Tx is plotted with respect to the change in distance in Figure 13. It is observed from Figure 13 that a significant improvement in RSS is achieved when asymmetric patch reflectarray antenna is used as a Tx. An average of −42 dBm of RSS is observed with asymmetric patch reflectarray antenna, which is almost 5 dBm higher than that of a standard horn antenna. This improvement in RSS is due to the higher gain of asymmetric patch reflectarray antenna as compared to a standard horn antenna. Moreover, the asymmetric patch reflectarray antenna offers almost the same RSS response with dual linear polarization operation.

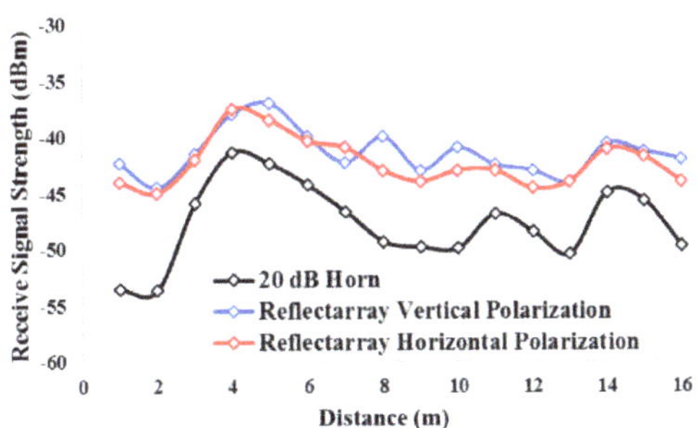

Figure 13. Comparison of measured RSS of asymmetric patch reflectarray antenna with standard horn antenna at a variable distance.

A higher RSS than a conventional horn antenna could also make device discovery easier in an urban microcell environment [31]. This measurement has also been repeated with the operating frequencies of 25 and 27 GHz, where a similar RSS response has been found. This analysis proves that, a good signal reception and detection at user equipment is expected if the proposed asymmetric patch reflectarray antenna is used as the 5G base station antenna. The only limitation associated with asymmetric patch reflectarray antenna is its narrow beamwidth, which tends to increase the number of sectors in a microcell. This limitation could be avoided by using a beamsteering strategy in the reflectarray antenna [6,19]. A wide reflection phase range, as in asymmetric patch reflectarray, would be beneficial for acquiring wide angle beamsteering operation.

5. Comparison of Asymmetric Patch Reflectarray Antenna with Other Related Works

The proposed design of asymmetric patch reflectarray antenna can also fall into the category of matasurface or frequency selective surface (FSS) due to its slotted ground structure. The latest technological advancement in metasurfaces with enhanced features are reported in [33,34]. The reported works of metasurfaces apply patch based dual layer technique to acquire beamsteering in reflection and transmission of the incident signals. Alternatively, this work proposes a single reflecting layer of asymmetric patches of ground ring slots for fixed beam operation. The two layer structure of metasurfaces are designed in such a way to improve the transmission of the incoming signals from the metasurface layer in both directions with minimum reflections. On the other hand, the proposed design of the reflectarray antenna offers maximum possible reflections from the patch based layer. The proposed asymmetric patch reflectarray antenna have ring slots in its ground plane, which could still pass some portion of the incident signals as transmission, but amount of this back radiating transmission would be very small as the reflected signal has a very high gain. Apart from some differences, the metasurfaces and the reflectarrays both use the same technique of patch based phase shift to control the direction of the transmitted or reflected signals. Additionally, if further layers are added in the proposed reflectarray design and its ground ring slots are widen further then it can be featured as a transmitive structure to offer features like a metasurface.

The proposed asymmetric patch reflectarray antenna is developed from the square patch reflectarray antenna. Asymmetry in the patch makes it to attain a wider reflection phase range and wider bandwidth performance as compared to an equivalent aperture square patch reflectarray antenna. In order to make a fair comparison of this work with other reflectarray antennas, an equivalent aperture square patch reflectarray antenna has been designed and experimentally tested. The square patch reflectarray antenna offers a measured 3 dB gain bandwidth of 6.9%, which is 4.6% lower than that of the asymmetric patch reflectarray antenna. The 5G research based on the reflectarray antenna is still very new and in its initial stages. There are only a few related and reputable works available in the literature that are summarized and compared with this work in Table 3. A most recent work proposing dual band reflectarray operation for 5G communication with dual linear polarization is reported in [35]. This dual band design uses two different elements for two different polarization and holds four elements in one unit cell for dual band operation. The asymmetric patch reflectarray offers a higher gain and wider bandwidth than this dual band design by using just a single patch element in one unit cell. A similar work proposed in [36] also uses the same technique for dual band operation, but with a single polarization. This work also has a narrower bandwidth performance than the asymmetric patch reflectarray antenna.

Table 3. Performance comparison of asymmetric patch reflectarray antenna with other related works.

Parameter	This Work	[35]	[4]	[36]	[19]	[37]	[38]
Technique	Asymmetric Patch	Dual band	–	Dual band	–	Transmitarray	Full Metallic
Frequency (GHz)	26	27/32	26	28/38	26	60	26
Patches in Unit Cell	1	4	1	4	1	4	1
Phase Span (°)	650	340/325	340	320/320	360	270	180
Array Elements	332	15 × 15	20 × 20	15 × 15	20 × 20	30 × 30	14 × 14
Array Size (λ2)	78.5	52.3/73	100	56.2/103	100	243.3	36.8
Gain (dB)	24.4	22.9/25.7	26.4	21/25.2	26.41	30.1	18.9
Max. Aperture Efficiency (%)	28	29/38	34	–	34	32.3	97.7 (Simulated)
Polarization	Dual Linear	Dual Linear	Single	Single	Single	Single	Single
Complexity	Low	Moderate	Low	Moderate	Low	High	High
1 dB Gain Bandwidth (%)	6.1	2.4/2.4	13.6	3.4/3.9	13.1	8.2	20 (1.5 dB) 4 (1 dB)
RSS Measurement	Yes	No	No	No	No	No	No

Moreover, the asymmetric patch reflectarray attains a wider reflection phase span with a lower design complexity than these two dual band designs. Two other similar works reported in [4] and [19] offer good bandwidth and gain performance with low design complexity, but for just a single polarization operation. Moreover, these two works present a reflectarray antenna with a conventional type of element, such as circular ring and rectangular patch, which offer a very narrow reflection phase range. In contrast, this work proposes a new asymmetric patch element with a wide reflection phase range for high-gain reflectarray operation with comparatively less number of elements. The asymmetric patch reflectarray antenna has also been compared with a transmitarray [37] and a full metallic reflectarray [38] that were proposed for 5G operation at 60 and 26 GHz respectively. Four different patches in five different layers make single polarized transmitarray an electrically large and very complicated structure as compared to the current work that operated at dual linear polarization. A full metallic design of a metallic reflectarray removes the dielectric losses and improves its efficiency performance. However, it still attains a lower gain and narrower bandwidth performance as compared to the asymmetric patch reflectarray antenna. This is because of a narrow reflection phase range of the unit cell of the metallic reflectarray antenna. On the other hand, the current work of the asymmetric patch reflectarray antenna has a wider reflection phase span with dual linear polarization operation. Moreover, this is the first time a reflectarray antenna is tested for RSS measurements with dual linear polarization operation for comparison with a standard horn antenna that is normally used for 5G based microcell measurements. Consequently, it can be said that the asymmetric patch reflectarray antenna proposed for 5G communication in this work, attains comparably better performance with low design complexity as compared to the related works of the published literature.

6. Conclusions

An asymmetric patch reflectarray unit cell element is proposed to increase the reflection phase range beyond its conventional range of 360°. A wide reflection phase span of 650° is acquired with an asymmetric patch of ground embedded circular ring slot. The bandwidth performance of the reflectarray antenna has been enhanced using asymmetric patch elements that also support high-gain operation with dual linear polarization. The limitation of high-loss performance due to a ground ring slot is avoided by acquiring a wide reflection phase range. On the other hand, this high-loss performance can be avoided by using a thicker substrate or by embedding an air gap between the substrate and the ground plane of the asymmetric patch element. The high cross polarization generated due to the asymmetry has been suppressed by mirroring the orientation of elements on the reflectarray surface. A maximum measured cross polarization of about −20 dB has been attained for both linear polarizations of the reflectarray antenna. This slightly high cross polarization is due to the single layer design with a single and geometrically asymmetric patch element. However, this value of cross polarization is still quite lower than the maximum side lobe level, and that is why it cannot affect the performance of the main lobe of the asymmetric patch reflectarray antenna. A circular aperture reflectarray antenna made of 332 asymmetric patch elements offers a 3 dB gain bandwidth of 3 GHz that is 11.5% at 26 GHz resonant frequency with a gain of 24.4 dB. The asymmetry in the patch element does not restrain it from supporting a dual linear polarization operation. Even a better bandwidth of 13.8% is acquired with a slight reduction of 0.5 dB in the gain performance for an orthogonal linear polarization operation. The proposed reflectarray antenna is also tested for indoor RSS measurements. In comparison with a standard horn antenna, it offers a better signal performance with an additional feature of polarization diversity in a microcell environment. A wider bandwidth performance, simple design, high-gain and dual linear polarization operation at high frequency, has made this proposed design of reflectarray suitable for 5G communications.

Author Contributions: Conceptualization, M.H.D. and M.H.J.; methodology, M.H.D. and M.I.A.; software, M.H.D., M.I.A. and A.Y.I.A.; validation, M.H.J., F.C.S. and M.R.K.; formal analysis, M.H.D. and M.I.A.; investigation, M.H.D. and O.H.; resources, M.H.J. and F.C.S.; data curation, M.H.D., A.Y.I.A. and O.H.; visualization, M.H.J. and M.R.K.; supervision, M.H.J. and F.C.S.; project administration, M.H.J. and M.R.K.; funding acquisition, M.H.J. and M.R.K. All authors have read and agreed to the published version of the manuscript.

Funding: This work is supported in part by the Ministry of Education Malaysia, Ministry of Science Technology and Innovation, Universiti Teknologi Malaysia and Research Management Center Universiti Tun Hussein Onn Malaysia under research fund E15501, Research Grant Vot 4J211, Vot 03G33, Vot 4S134, and Vot13H26.

Acknowledgments: Authors wish to thank the staff of Wireless Communication Centre, Universiti Teknologi Malaysia for the technical support.

Conflicts of Interest: The authors declare no conflict of interest.

References

1. Boccardi, F.; Heath, R.; Lozano, A.; Marzetta, T.L.; Popovski, P. Five disruptive technology directions for 5G. *IEEE Commun. Mag.* **2014**, *52*, 74–80. [CrossRef]
2. Vook, F.W.; Ghosh, A.; Thomas, T.A. MIMO and beamforming solutions for 5G technology. In Proceedings of the 2014 IEEE MTT-S International Microwave Symposium (IMS2014), Tampa, FL, USA, 1–6 June 2014; pp. 1–4. [CrossRef]
3. Dahri, M.H.; Jamaluddin, M.H.; Inam, M.; Kamarudin, M.R. A Review of Wideband Reflectarray Antennas for 5G Communication Systems. *IEEE Access* **2017**, *5*, 17803–17815. [CrossRef]
4. Inam, M.; Dahri, M.H.; Jamaluddin, M.H.; Seman, N.; Kamarudin, M.R.; Sulaiman, N.H. Design and characterization of millimeter wave planar reflectarray antenna for 5G communication systems. *Int. J. RF Microw. Comput. Eng.* **2019**, *29*. [CrossRef]
5. Dahri, M.H.; Inam, M.; Jamaluddin, M.H.; Kamarudin, M.R. A Review of High Gain and High Efficiency Reflectarrays for 5G Communications. *IEEE Access* **2017**. [CrossRef]
6. Hashim Dahri, M.; Jamaluddin, M.H.; Khalily, M.; Abbasi, M.I.; Selvaraju, R.; Kamarudin, M.R. Polarization Diversity and Adaptive Beamsteering for 5G Reflectarrays: A Review. *IEEE Access* **2018**, *6*, 19451–19464. [CrossRef]

7. ITU Final Acts WRC-15. In Proceedings of the World Radiocommunication Conference, Geneva, Switzerland, 2–27 November 2015; Available online: http://www.itu.int/pub/R-ACT-WRC.12-2015 (accessed on 2 September 2020).

8. Andrews, J.G.; Buzzi, S.; Choi, W.; Hanly, S.V.; Lozano, A.; Soong, A.C.K.; Zhang, J.C. What will 5G be? *IEEE J. Sel. Areas Commun.* **2014**, *32*, 1065–1082. [CrossRef]

9. Greco, F.; Boccia, L.; Arnieri, E. A Ka-Band Cylindrical Paneled Reflectarray Antenna. *Electronics* **2019**, *8*, 654. [CrossRef]

10. Huang, J.; Encinar, J. *Reflectarray Antennas*; Wiley-IEEE Press: Hoboken, NJ, USA, 2007.

11. Huang, J. Analysis of microstrip reflectarray antenna for microspacecraft applications. *Telecommun. Data Acquis. Prog. Rep.* **1994**, 153–173, Bibcode: 1994TDAPR.120..153H.

12. Fazaelifar, M.; Jam, S.; Basiri, R. Design and fabrication of a wideband reflectarray antenna in Ku and K bands. *AEU Int. J. Electron. Commun.* **2018**, *95*, 304–312. [CrossRef]

13. Chen, G.T.; Jiao, Y.C.; Zhao, G. A Reflectarray for Generating Wideband Circularly Polarized Orbital Angular Momentum Vortex Wave. *IEEE Antennas Wirel. Propag. Lett.* **2019**, *18*, 182–186. [CrossRef]

14. Karimipour, M.; Aryanian, I. Demonstration of Broadband Reflectarray Using Unit Cells with Spline-Shaped Geometry. *IEEE Trans. Antennas Propag.* **2019**, *67*, 3831–3838. [CrossRef]

15. Liu, Y.; Cheng, Y.J.; Lei, X.Y.; Kou, P.F. Millimeter-Wave Single-Layer Wideband High-Gain Reflectarray Antenna with Ability of Spatial Dispersion Compensation. *IEEE Trans. Antennas Propag.* **2018**, *66*, 6862–6868. [CrossRef]

16. Xue, F.; Wang, H.-J.; Yi, M.; Liu, G. A broadband KU-band microstrip reflectarray antenna using single-layer fractal elements. *Microw. Opt. Technol. Lett.* **2016**, *58*, 658–662. [CrossRef]

17. Derafshi, I.; Komjani, N.; Mohammadirad, M. A single-layer broadband reflectarray antenna by using quasi-spiral phase delay line. *IEEE Antennas Wirel. Propag. Lett.* **2015**, *14*, 84–87. [CrossRef]

18. Bhardwaj, S.; Rahmat-samii, Y. Revisiting the Generation of Cross- Polarization in Rectangular Patch Antennas: A Near-Field Approach. *IEEE Antennas Propag. Mag.* **2014**, *56*, 14–38. [CrossRef]

19. Abbasi, M.I.; Dahri, M.H.; Jamaluddin, M.H.; Seman, N.; Kamarudin, M.R.; Sulaiman, N.H. Millimeter Wave Beam Steering Reflectarray Antenna Based on Mechanical Rotation of Array. *IEEE Access* **2019**, *7*, 145685–145691. [CrossRef]

20. Balanis, C.A. *Antenna; Theory Analysis and Design*, 3rd ed.; John Wiley and Sons: Hoboken, NJ, USA, 2005.

21. Rajagopalan, H.; Rahmat-Samii, Y. On the reflection characteristics of a reflectarray element with low-loss and high-loss substrates. *IEEE Antennas Propag. Mag.* **2010**, *52*, 73–89. [CrossRef]

22. Pozar, D.M.; Metzler, T.A. Analysis of a reflectarray antenna using microstrip patches of variable size. *Electron. Lett.* **1993**, *29*, 657–658. [CrossRef]

23. Florencio, R.; Martínez-De-rioja, D.; Martínez-De-rioja, E.; Encinar, J.A.; Boix, R.R.; Losada, V. Design of ku-and ka-band flat dual circular polarized reflectarrays by combining variable rotation technique and element size variation. *Electronics* **2020**, *9*, 985. [CrossRef]

24. Chaharmir, M.R.; Shaker, J.; Cuhaci, M.; Sebak, A. Reflectarray with variable slots on ground plane. *IEE Proc. Microw. Antennas Propag.* **2003**, *150*, 18–21. [CrossRef]

25. Chen, P.; Yang, X.D.; Chen, C.Y.; Ma, Z.H. Broadband Multilayered Array Antenna with EBG Reflector. *Int. J. Antennas Propag.* **2013**, *2013*, 1–5. [CrossRef]

26. Malfajani, R.S.; Atlasbaf, Z. Design and Implementation of a Dual-Band Single Layer Reflectarray in X and K Bands. *IEEE Trans. Antennas Propag.* **2014**, *62*, 4425–4431. [CrossRef]

27. Inam, M.; Ismail, M.Y. Reflection loss and bandwidth performance of X-band infinite reflectarrays: Simulations and measurements. *Microw. Opt. Technol. Lett.* **2011**, *53*, 77–80.

28. Pozar, D.M.; Targoski, S.D.; Syrigos, H.D. Design of millimeter wave microstrip reflectarrays. *IEEE Trans. Antennas Propag.* **1997**, *45*, 287–296. [CrossRef]

29. Rengarajan, S.R. Reciprocity considerations in microstrip reflectarrays. *IEEE Antennas Wirel. Propag. Lett.* **2009**, *8*, 1206–1209. [CrossRef]

30. Kumar, C.; Guha, D. Asymmetric Geometry of Defected Ground Structure for Rectangular Microstrip: A New Approach to Reduce its Cross-Polarized Fields. *IEEE Trans. Antennas Propag.* **2016**, *64*, 6–9. [CrossRef]

31. Hayat, O.; Ngah, R.; Hashim, S.Z.M. Sector Scanning Algorithm (SSA) for Device Discovery in D2D Communication. *Int. J. Electron.* **2020**, *1*, 1–15. [CrossRef]

32. Maccartney, G.R.; Rappaport, T.S.; Ghosh, A. Base Station Diversity Propagation Measurements at 73 GHz Millimeter-Wave for 5G Coordinated Multipoint (CoMP) Analysis. In Proceedings of the 2017 IEEE Globecom Workshops (GC Wkshps), Singapore, 4–8 December 2017; pp. 1–7.
33. Taravati, S.; Eleftheriades, G.V. Full-Duplex Nonreciprocal Beam Steering by Time-Modulated Phase-Gradient Metasurfaces. *Phys. Rev. Appl.* **2020**, *14*, 014027. [CrossRef]
34. Taravati, S.; Khan, B.A.; Gupta, S.; Achouri, K.; Caloz, C. Nonreciprocal Nongyrotropic Magnetless Metasurface. *IEEE Trans. Antennas Propag.* **2017**, *65*, 3589–3597. [CrossRef]
35. Costanzo, S.; Member, S.; Venneri, F.; Borgia, A.; Massa, G.D.I.; Member, L.S. Dual-Band Dual-Linear Polarization Reflectarray for mmWaves/5G Applications. *IEEE Access* **2020**, *8*, 78183–78192. [CrossRef]
36. Costanzo, S.; Venneri, F.; Borgia, A.; Massa, G. Di A Single-Layer Dual-Band Reflectarray Cell for 5G Communication Systems. *Int. J. Antennas Propag.* **2019**, *2019*, 8–11. [CrossRef]
37. Erfani, E.; Safavi-Naeini, S.; Tatu, S. Design and analysis of a millimetre-wave high gain antenna. *IET Microw. Antennas Propag.* **2019**, *13*, 1586–1592. [CrossRef]
38. Mei, P.; Zhang, S.; Pedersen, G.F. A Low-Cost, High-Efficiency and Full-Metal Reflectarray Antenna with Mechanically 2-D Beam-Steerable Capabilities for 5G Applications. *IEEE Trans. Antennas Propag.* **2020**, 1–10. [CrossRef]

 *electronics*

Article

Isolation Improvement in UWB-MIMO Antenna System Using Slotted Stub

Ahsan Altaf [1], Amjad Iqbal [2,3,*], Amor Smida [4,5,*], Jamel Smida [6], Ayman A. Althuwayb [7], Saad Hassan Kiani [8], Mohammad Alibakhshikenari [9,*] and Francisco Falcone [10,11] and Ernesto Limiti [9]

[1] Electrical Engineering Department, Istanbul Medipol University, Istanbul 34083, Turkey; aaltaf@st.medipol.edu.tr

[2] Centre For Wireless Technology, Faculty of Engineering, Multimedia University, Cyberjaya 63100, Malaysia

[3] Electrical Engineering Department, CECOS University of IT and Emerging Sciences, Peshawar 25000, Pakistan

[4] Department of Medical Equipment Technology, College of Applied Medical Sciences, Majmaah University, AlMajmaah 11952, Saudi Arabia

[5] Microwave Electronics Research Laboratory, Department of Physics Faculty of Mathematical, Physical and Natural Sciences of Tunis, Tunis ElManar University, Tunis 2092, Tunisia

[6] College of Applied Sciences, AlMaarefa University, Riyadh 11952, Saudi Arabia; jsmida@mcst.edu.sa

[7] Electrical Engineering Department, Jouf University, Sakaka, Aljouf 72388, Saudi Arabia; aaalthuwayb@ju.edu.sa

[8] Department of Electrical Engineering, City University of Science and Information Technology, Peshawar 25000, Pakistan; iam.kiani91@gmail.com

[9] Electronic Engineering Department, University of Rome "Tor Vergata", Via del Politecnico 1, 00133 Rome, Italy; limiti@ing.uniroma2.it

[10] Electrical Engineering Department, Public University of Navara, 31006 Pamplona, Spain; francisco.falcone@unavarra.es

[11] Institute of Smart Cities, Public University of Navarre, 31006 Pamplona, Spain

* Correspondence: amjad730@gmail.com (A.I.); a.smida@mu.edu.sa (A.S.); Alibakhshikenari@ing.uniroma2.it (M.A.); Tel.: +60-1-128-78-4475 (A.I.)

Received: 11 September 2020; Accepted: 25 September 2020; Published: 27 September 2020

Abstract: Multiple-input multiple-output (MIMO) scheme refers to the technology where more than one antenna is used for transmitting and receiving the information packets. It enhances the channel capacity without more power. The available space in the modern compact devices is limited and MIMO antenna elements need to be placed closely. The closely spaced antennas undergo an undesirable coupling, which deteriorates the antenna parameters. In this paper, an ultra wide-band (UWB) MIMO antenna system with an improved isolation is presented. The system has a wide bandwidth range from 2–13.7 GHz. The antenna elements are closely placed with an edge to edge distance of 3 mm. In addition to the UWB attribute of the system, the mutual coupling between the antennas is reduced by using slotted stub. The isolation is improved and is below -20 dB within the whole operating range. By introducing the decoupling network, the key performance parameters of the antenna are not affected. The system is designed on an inexpensive and easily available FR-4 substrate. To better understand the working of the proposed system, the equivalent circuit model is also presented. To model the proposed system accurately, different radiating modes and inter-mode coupling is considered and modeled. The EM model, circuit model, and the measured results are in good agreement. Different key performance parameters of the system and the antenna element such as envelope correlation coefficient (ECC), diversity gain, channel capcity loss (CCL) gain, radiation patterns, surface currents, and scattering parameters are presented. State-of-the-art comparison with the recent literature shows that the proposed antenna has minimal dimensions, a large bandwidth, an adequate gain value and a high isolation. It is worth noticeable that the proposed antenna has high isolation even the patches has low edge-to-edge gap (3 mm). Based on its good performance and compact dimensions, the proposed antenna is a suitable choice for high throughput compact UWB transceivers.

Keywords: isolation enhancement; surface waves; gain; circuit model; slotted-stub; MIMO antennas

1. Introduction

In the last decade, mobile technology enables researchers to find solutions for systems consist of multiple devices, sensors, and components [1–5]. Antennas are one of the important components of these multi-device systems [6–8]. With MIMO technology, need for antennas that can operate over a wide range of frequencies is inevitable [9–12]. Ultra wide band (UWB) antennas have better channel capacity, envelope correlation, and diversity gain as compared to traditional narrowband antennas [13].

According to Federal Communication Commission (FCC) protocol, the UWB ranging from 3.2 to 10.6 GHz covering indoor applications and handheld devices [14]. Several designs have been proposed and researched in [15–19] for UWB services. In [15] a V shape monopole antenna with staircase shaped defected ground structure is proposed having realized gain of 4 dB with size of 25 × 26 mm for UWB applications. The Multiple Input Multiple Output (MIMO) technology offers higher data rates with better propagation response in multipath propagating environment and noise immunity [20–22]. Enhanced Isolation is desirable in MIMO antenna systems as the lower coupling ensures the better antenna performance characteristics [23,24]. In order to suppress surface waves of MIMO radiating elements the use of defected ground structures (DGSs) [25,26], electromagnetic band gap structures (EBGs) [27,28], and line resonators [29] are commonly used.

In [30] an UWB MIMO antenna having dual notch response is investigated by using several metal strips with modified ground plane to reduce mutual coupling between the radiating elements. In [31] an UWB MIMO antenna with compact size (25 × 40 mm^2) and polarization diversity is presented. The design is complex because of two different ground planes. In [32] a bio inspired leaf shaped antenna is presented for wideband and sensing applications, however with a large dimensional size of 314 × 121 mm. The design is not suitable for MIMO configuration for handheld devices. The use of black carbon film on the UWB MIMO reduced the coupling effects in [33] due to absorption losses, however the cost of the system with such an approach increases.

In this paper, an UWB MIMO antenna system for 2–13.7 GHz is presented. The system is composed of two radiating elements. The mutual coupling between the antennas is reduced by using slotted stubs. The isolation is improved and is below −20 dB for the whole operating range. The system is fabricated on an easily available, inexpensive, and widely used FR4 substrate. The system is designed using a EM full-wave commercial software HFSS. A detailed study regarding the antenna design, decoupling network, and other parameters of the design is presented. The circuit model is presented to verify the EM model. A prototype is fabricated to verify the simulated results. The comaprison between the simulated and measured results is presented and it is found that they are in good agreement. The effects of decoupling network over the performance of the antennas and the MIMO system is also studied. It is found that ECC and Diversity Gain (DG) for the proposed system are 0.15 and 9.85 dBi respectively. The channel capcity loss (CCL) is less than 0.06 bps/Hz for the whole operating bandwidth.

2. Design Methodology

The design and prototype models of the proposed UWB MIMO antenna system are presented in Figure 1. The system is designed on an FR4 substrate with a thickness of 1.6 mm. The dielectric constant of the substrate is assumed to be 4.4. The top side includes two radiating stepped-shaped elements with a 50 Ω feed line. The stepped shape is designed to increase the path for the flow of current. By increasing the path for the current, a wide bandwidth is achieved. The two rectangles are designed at the top of each element to achieve wide-band matching. A rectangular ground plane with a rectangular slot is designed at the back side of the substrate. The antenna elements share a common ground plane. The upper edge of the common ground is stepped and incorporated with the plus-shaped slotted stubs to improve isolation between the antenna elements. The stubs act as a bandstop filter for the desired frequency range. It produces transmission zeros within the radiating elements that in turn disturb the flow of current, surface waves, and near fields. The dimensions of the antenna elements, slotted stubs, ground plane, and feed lines are shown in Figure 1.

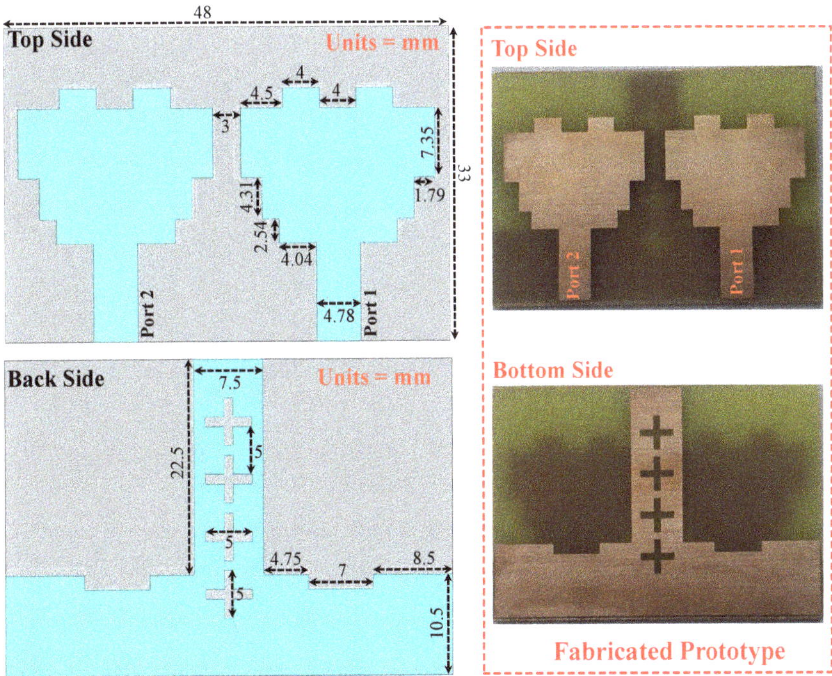

Figure 1. Top and bottom view of the proposed ultra wide-band multiple input multiple output (UWB-MIMO) antenna.

2.1. Single Unit Antenna

A detailed evolution procedure for the proposed setup is presented in Figure 2. In Step 1, a stepped-shaped single element with a slotted ground plane is designed. The ground plane covers the area between the lower edge and the feed line. In Step-2, two radiating elements with a common ground plane is designed. In other words, step-1 is replicated along the *x*-axis to get a MIMO antenna system. The edge to edge distance between the antenna elements is 3 mm. In the final step, slotted stubs are incorporated at the back side within the antenna elements to reduce the mutual coupling between the radiating elements. The slotted stubs suppress the surface waves and the near fields reducing the coupling between the antennas keeping it below -20 dB for the whole desired frequency range. The step-by-step evolution of the proposed setup is shown in Figure 2. To understand the proposed MIMO antenna system, an equivalent circuit model for a single antenna using lumped elements is presented. In Figure 3 the transmission line is modeled as an impedance transformer with the coupling ratio of X1:1 in the circuit model. Since patch antenna is a resonating circuit, one can use circuit theory to model the radiating element approximately by the RLC circuit [34]. Where R is the radiation resistance of the radiating mode of the antenna, and L and C describe the resonant circuit that is responsible for the desired resonant frequency. The proposed antenna setup is a wide-band, therefore the desired response is a combination of different radiating modes. In addition to the combination of these radiating modes, the coupling between the modes should also be included to model the system accurately. The equivalent circuit model for a single element is shown in the inset of Figure 3. The radiating modes are modeled by RLC combinations while the two LC (tank-circuits) combinations represent inter-mode coupling [35,36]. The circuit model is studied in Keysight Advance Design System (ADS). The values of each component and the comparison of S-parameters between the EM and the circuit model are depicted in Figure 3. The results are in good agreement.

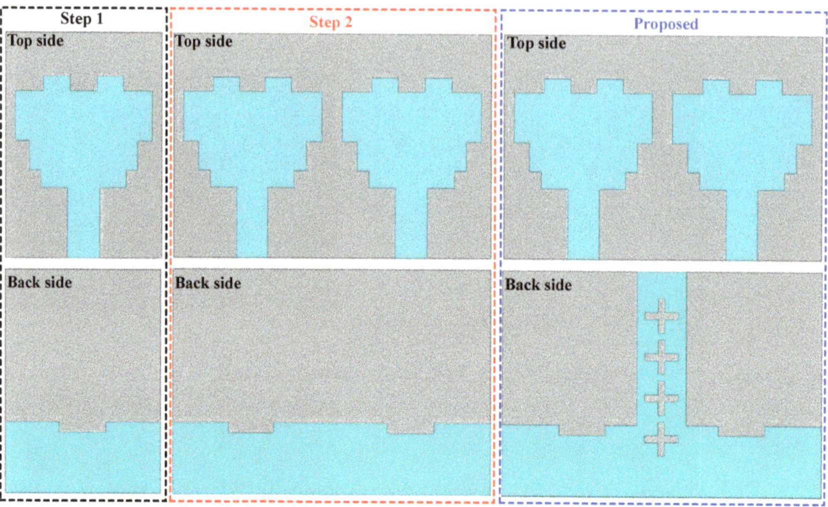

Figure 2. Design evolution stages of the proposed UWB-MIMO antenna.

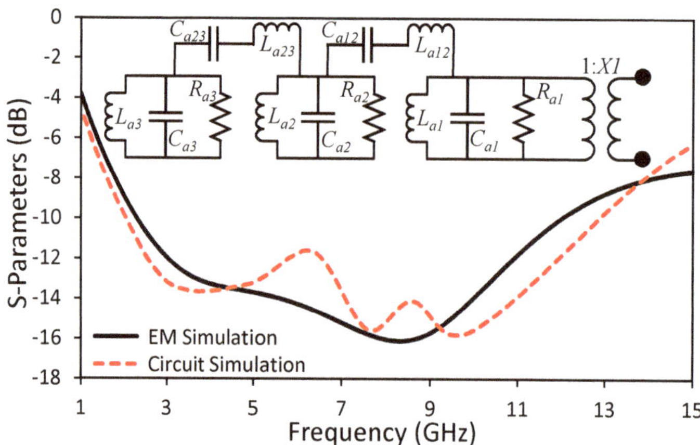

Figure 3. Lumped element equivalent circuit model (R_{a1} = 295 Ω, R_{a2} = 54.45 Ω, R_{a3} = 60 Ω, C_{a1} = 3.45 pF, C_{a2} = 3.82 pF, C_{a3} = 4.8 pF, L_{a1} = 1.55 nH, L_{a2} = 7.42 nH, L_{a3} = 3.85 nH, L_{a12} = 0.6 nH, C_{a12} = 1.5 pF, L_{a23} = 0.7 nH and C_{a23} = 1.22 pF) and comparison of EM and circuit model results.

2.2. UWB-MIMO Antenna System without Slotted Stub

In this section, the circuit model for the proposed MIMO antenna system is presented. As discussed in the previous section, each antenna element can be modeled as a combination of the RLC circuit, and a feed line can be modeled as an impedance transformer. Since our antennas exhibit wide-band response, many radiating modes are contributing. In other words, a combination of radiating modes makes it a wide-band antenna. From circuit theory, it is evident that inter-mode coupling affects the response of the circuit. Therefore, to model the MIMO antenna system accurately, one should not only consider the radiating modes but also the inter-mode coupling. In this paper, for simplicity, three radiating modes and inter-mode coupling between antennas are considered for corresponding modes only. For instance, the coupling between mode-1 of antenna-1 with mode-1 of antenna-2 is considered but the coupling of mode-1 of antenna-1 with the coupling of mode-2 of antenna-2 is not discussed. The equivalent circuit model of the proposed system without slotted stubs is shown Figure 4. The feed lines of the radiating elements are modeled by impedance transformers with a coupling ratio of X1:1. Each antenna element radiating modes are modeled by a combination of RLC circuits while the tank circuits represent the inter-mode couplings. The mutual coupling between the corresponding radiating mode of antenna elements is represented by M_1, M_2, and M_3. The circuit model is designed in ADS. The comparison for S-parameters results between the EM and the circuit model and the values of each component are shown Figure 5. It can be seen from Figure 5, the results are in a good agreement for the desired frequency band.

Component	Value
R_{a1}	1546 Ω
R_{a2}	1861 Ω
R_{a3}	47 Ω
C_{a1}	0.5 pF
C_{a2}	0.5 pF
C_{a3}	10.45 pF
L_{a1}	0.5 nH
L_{a2}	0.5 nH
L_{a3}	0.93 nH
L_{a12}	19.4 nH
L_{a23}	9.6 nH
C_{a12}	20 pF
C_{a23}	3.23 pF
$X1$	0.92
M_1	1.575
M_2	1.22
M_3	1.02

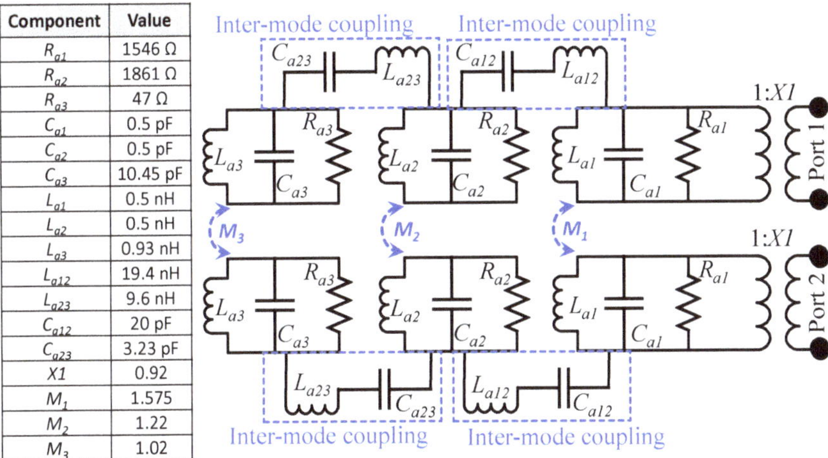

Figure 4. Lumped element equivalent circuit model of the UWB-MIMO system without decoupling structure.

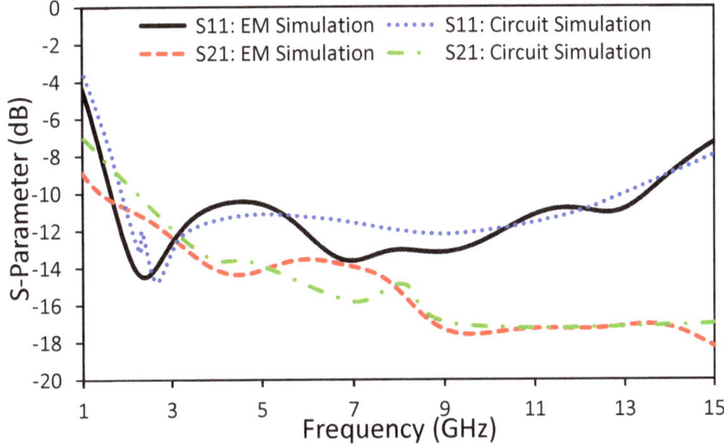

Figure 5. Lumped element equivalent circuit simulation comparison with EM simulation without decoupling structure.

2.3. UWB-MIMO Antenna System with Slotted Stub

In this section, the equivalent circuit model for slotted stubs is presented. The slotted stubs are added within the antenna elements at the ground plane side to reduce the mutual coupling between the antennas, suppress the surface waves and near fields. To model accurately the proposed MIMO antenna system with the decoupling structure, the same assumptions regarding the radiating modes and inter-mode coupling are considered. In other words, three radiating modes and corresponding inter-mode couplings are modeled for the radaiting structures. The radiating modes are modeled by the combination of RLC circuits, transmission lines are modeled by impedance transformers, and inter-mode coupling is modeled by LC-tank circuits. Since the mutual coupling can be reduced by a band stop response by introducing transmission zeros within the operating range. The decoupling structure is modeled by the combination of RLC circuits. These RLC circuits are coupled with the radiating modes by M_{1dc}, M_{2dc}, and M_{3dc} coupling ratios. By using the coupling structure, the isolation for the proposed system is well below 20 dB for the whole bandwidth. The value of each component and the equivalent circuit model are depicted in Figure 6. The simulated results of the EM model and the circuit model are in good agreement, as shown in Figure 7.

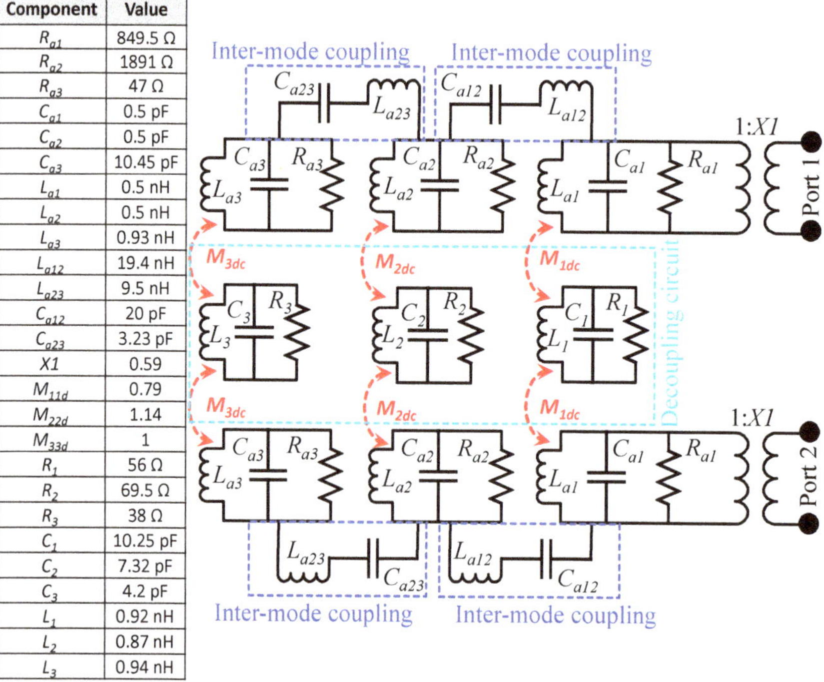

Component	Value
R_{a1}	849.5 Ω
R_{a2}	1891 Ω
R_{a3}	47 Ω
C_{a1}	0.5 pF
C_{a2}	0.5 pF
C_{a3}	10.45 pF
L_{a1}	0.5 nH
L_{a2}	0.5 nH
L_{a3}	0.93 nH
L_{a12}	19.4 nH
L_{a23}	9.5 nH
C_{a12}	20 pF
C_{a23}	3.23 pF
$X1$	0.59
M_{11d}	0.79
M_{22d}	1.14
M_{33d}	1
R_1	56 Ω
R_2	69.5 Ω
R_3	38 Ω
C_1	10.25 pF
C_2	7.32 pF
C_3	4.2 pF
L_1	0.92 nH
L_2	0.87 nH
L_3	0.94 nH

Figure 6. Lumped element equivalent circuit model of the UWB-MIMO system with decoupling structure.

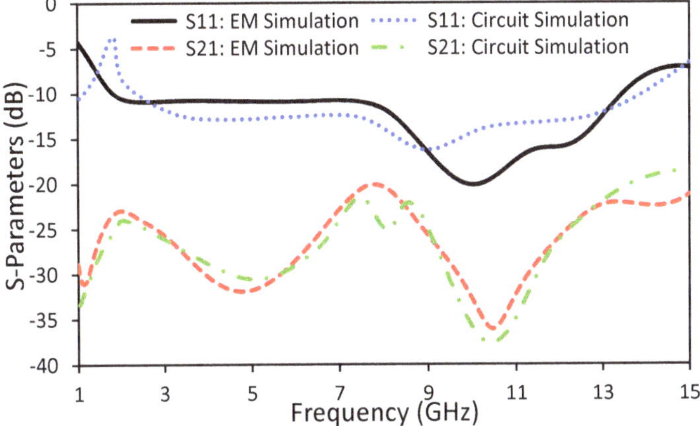

Figure 7. Lumped element equivalent circuit simulation comparison with EM simulation in the presence of decoupling structure.

3. Results and Discussion

The proposed MIMO antenna system with an isolation network was fabricated using the LPKF machine. The system was measured using a two-port vector network analyzer. The radiation patterns were measured in an anechoic chamber. By introducing the decoupling network, key performance parameters of the antenna were not affected. The simulated and measured S-parameters results for the proposed system are shown in Figure 8. The results were in good agreement. It can be seen that the isolation between the radiating elements was below −20 dB within the whole operating range.

The slotted stubs were used to improve isolation between the antennas. When there was no decoupling network, the surface waves were coupled to the non-excited radiating element resulting in high mutual coupling. On the other hand, when the decoupling structure was incorporated within the elements, one can see that the surface waves and near fields were suppressed in the non-excited antenna. The surface waves and the near fields were concentrated within the decoupling structure resulting in high isolation between the antenna elements. This phenomenon can be observed in Figure 9. The surface current distribution with and without the slotted stubs at 10 GHz is shown in Figure 9.

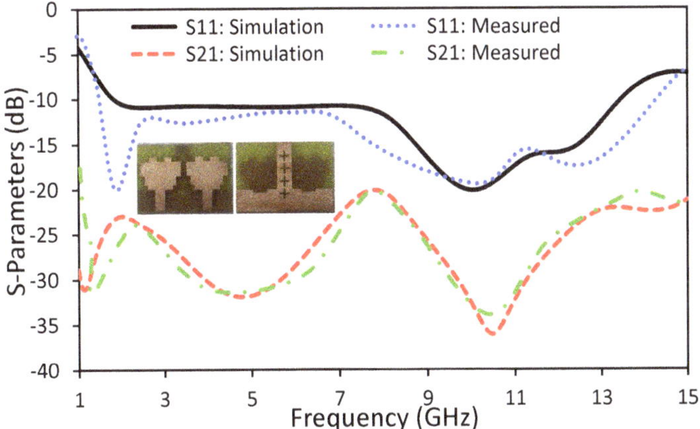

Figure 8. Simulated and measured S-parameters of the antenna.

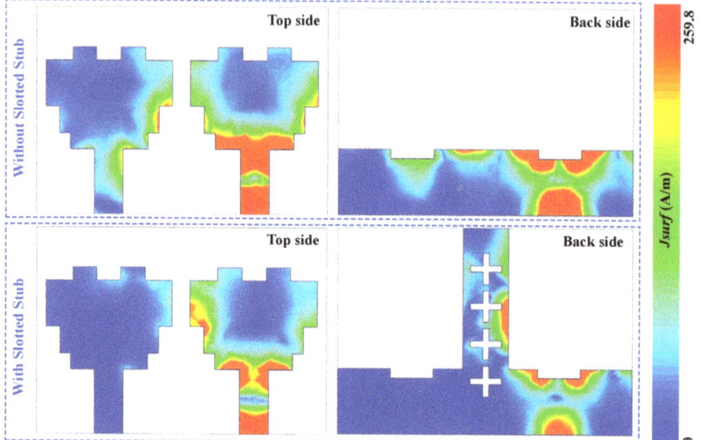

Figure 9. Surface current distribution with and without slotted stub at 10 GHz.

Gain is one of the key parameter of the antenna. The simulated and measured gain of the antenna is greater than 1.1 dBi with a maximum value of 4.3 dBi at 12 GHz. The simulated and measured gains are in good agreement for the whole operating bandwidth as shown in Figure 10. The far-field patterns of the proposed system is shown in Figure 11. The radiation patterns are shown for three different frequencies, i.e., 2.4, 5, and 10 GHz. The comparison between the simulated and measured xz- and yz-plane radiation characteristics are shown in Figure 11. The results were in good agreement.

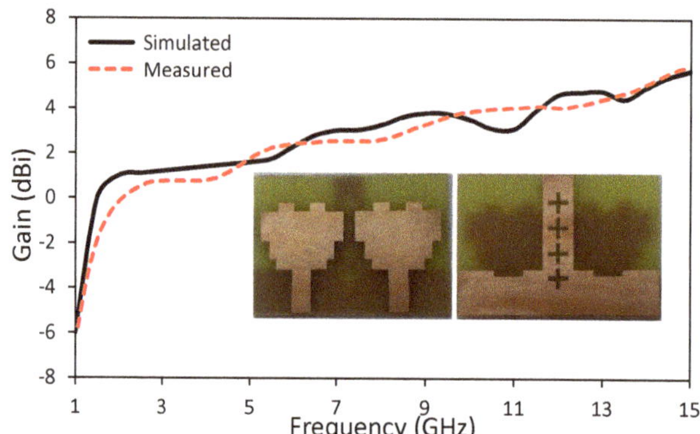

Figure 10. Simulated and measured peak gain of the antenna.

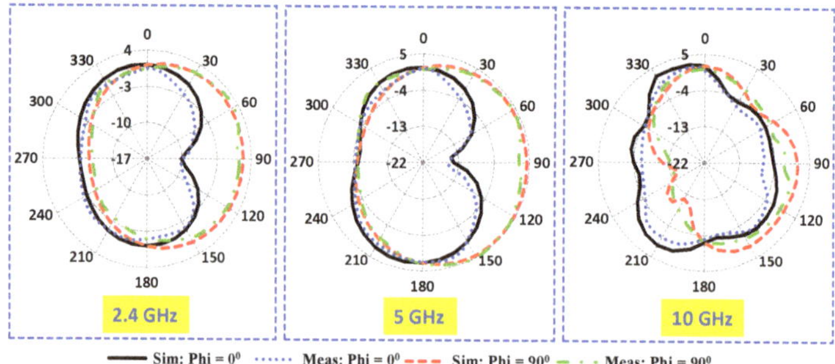

Figure 11. Radiation pattern of the antenna at 2.4, 5 and 10 GHz.

4. MIMO Parameters

In this section, key performance parameters of a MIMO system are discussed. Envelope Correlation Coefficient (ECC) and Diversity Gain (DG) are one of the vital parameters of the MIMO system. ECC defines the effect of one antenna over the performance of the other antenna within a multiple antenna system. In other words, it measures how a member is affecting the performance of the other member within a system. ECC can be calculated either by using the S-parameters or radiation characteristics as represented in Equations (1) and (2) respectively. In this paper, ECC is calculated using Equation (2).

$$ECC = \frac{|S_{11}^{*}S_{12} + S_{22}^{*}S_{21}|^2}{\left[1 - (|S_{11}|^2) + |S_{12}|^2\right]\left[1 - (|S_{22}|^2) + |S_{21}|^2\right]} \tag{1}$$

$$ECC = \frac{|\int\int_{4\pi}(\mathbf{B}_i(\theta,\phi)) \times \mathbf{B}_j(\theta,\phi))d\Omega|^2}{\int\int_{4\pi}|\mathbf{B}_i(\theta,\phi))|^2d\Omega \int\int_{4\pi}|\mathbf{B}_j(\theta,\phi))|^2d\Omega} \tag{2}$$

where S_{11}/S_{22} and S_{21}/S_{12} are the reflection and transmission coefficient of the antenna. $\mathbf{B}_i(\theta,\phi)$ is the three dimensional radiation pattern upon excitation of the i-th antenna and $\mathbf{B}_j(\theta,\phi)$ is the three dimensional radiation pattern upon excitation of the j-th antenna, and Ω represents the solid angle.

On the other hand, the DG is the selection of the strongest signal from N number of signals. It is calculated through the following equation.

$$DG = 10\sqrt{1 - (ECC)^2} \tag{3}$$

The simulated ECC and DG for the proposed system is shown in Figure 12. It can be seen that the DG was more than 9.85 dBi and the ECC was less than 0.15 within the band of interest. One of the important MIMO system parameters is Channel Capacity Loss (CCL). The measured and simulated CCL of the proposed system is depicted in Figure 13. It is observed that CCL was less than 0.06 bps/Hz for the whole operating bandwidth.

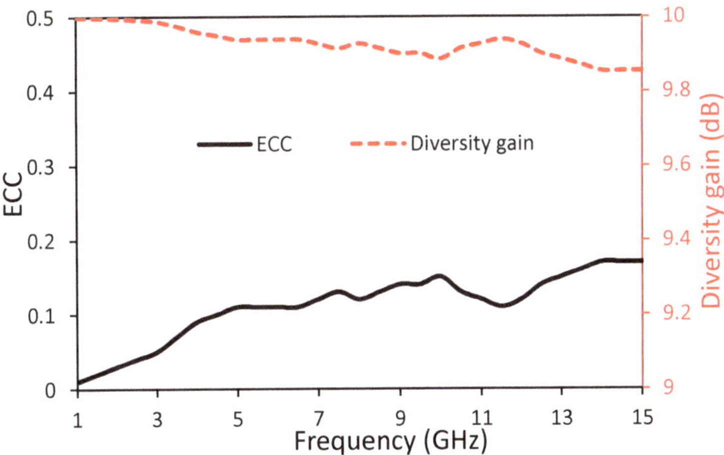

Figure 12. Envelope correlation coefficient (ECC) and Diversity Gain (DG) of the UWB-MIMO antenna system.

A detailed comparison between the proposed and published UWB-MIMO antenna is tabulated in Table 1. It is noticeable that the proposed antenna cover minimal dimensions, has large bandwidth, adequate gain value and high isolation. It is worth noticeable that the proposed antenna has high isolation even the patches has low edge-to-edge gap (3 mm). Moreover, we extracted circuit model in each step. The isolation and bandwidth enhancements phenomena is discussed using circuit theory.

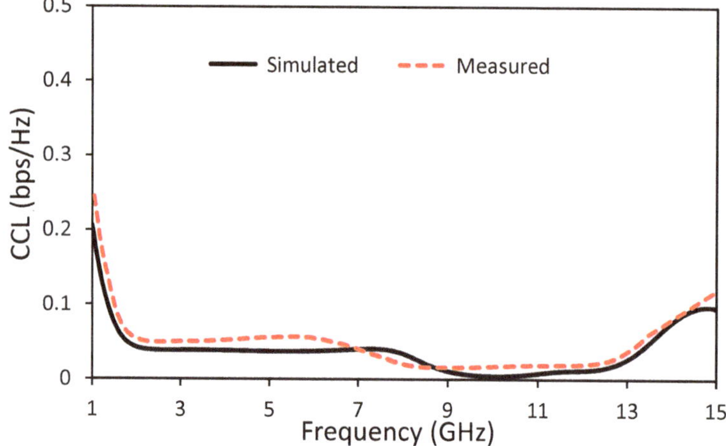

Figure 13. Simulated and measured CCL of the UWB-MIMO antenna system.

Table 1. Performance comparison with UWB-MIMO antennas.

Ref.	Size (mm²)	Edge-to-Edge Gap (mm)	Operating Frequency Range (GHz)	Isolation (dB)	Gain (dBi)	Common Ground	Circuit Model
[6]	40 × 80	NG	4.5–8	>25	2–4	No	No
[7]	27 × 47	NG	3.1–10.6	>20	2.8–5.4	Yes	No
[8]	18 × 36	21.5	3–40	>15	0–9	Yes	No
[9]	68 × 35	NG	3.1–10.6	20	1.7–4.2	No	No
[10]	81 × 81	NG	3.1–11.14	>18	4–8.48	No	No
[11]	37 × 45	3	3.1–5	>20	>−2	Yes	No
[12]	40 × 68	34	3.2–10.6	>15	3–5.5	Yes	No
[13]	50 × 30	NG	2.5–14.5	>20	0.3–4.3	Yes	No
This Work	**33 × 48**	**3**	**2–13.7**	**>20**	**1.1–4.3**	**Yes**	**Yes**

Designing a UWB-MIMO antenna has many challenges. The main challenges include achieving a wide bandwidth with omnidirectional patterns, high gain and acceptable isolation within radiating elements. Moreover, it is necessary to maintain the parameters of antenna before and after the decoupling structure. In this work, the isolation between the antennas is improved by etching slots in the ground plane and omnidirectional patterns are achieved for the whole frequency range. The proposed antenna cannot be used for the base station because the base station requires high gain antennas. In other words, this antenna is suitable for mobile devices due to its miniature nature and omnidirectional patterns. One of the limitations is that the phase and magnitude of the radiated fields are frequency dependent. UWB antennas are a combination of different resonances within a frequency range. The impedance, surface waves, and radiated fields vary and affect the performance of the antenna. This makes the response of the antenna unpredictable. By deploying different techniques like DGS, EBG, and slots within the antenna and/or ground plane, the response can be controlled.

5. Conclusions

In this article, an UWB MIMO antenna system with an isolation of −20 dB is presented. The UWB response is achieved by designing a stepped-shaped antenna while the isolation is improved by using slotted stubs etched in the ground plane within the radiating elements. The system is fabricated on an FR4 substrate. The circuit theory is used to model the circuit model and compared the results with the

EM model. The simulated design is fabricated and key performance parameters are calculated and measured. The simulated, measured, and the circuit model results are in good agreement. It is found that ECC and DG for the proposed system are 0.15 and 9.85 dBi respectively. The CCL is less than 0.06 bps/Hz for the whole operating bandwidth. Future work will focus on increasing the number of transmitting and receiving antennas to enhance the throughput. Moreover, the coupling effect between the antenna elements can be enhanced by combining the proposed method with any other method. In fact, such a decoupling method will be helpful to develop universal guidelines for compact UWB-MIMO antenna systems.

Author Contributions: design and concept, A.A. and A.I.; methodology, A.A. A.I.; investigation, A.I. and S.H.K.; resources, A.I.; writing—original draft preparation, A.A., A.I., A.S., J.S. and S.H.K.; writing—review and editing, A.A., A.I., A.S., J.S., A.A.A. and S.H.K.; validation, A.A., A.I., A.S., J.S., A.A.A., S.H.K. M.A., E.L. and F.F.; supervision, A.A., A.I., A.S., J.S., A.A.A., S.H.K. M.A., E.L. and F.F.; project administration, A.A., A.I., A.S., J.S., A.A.A., S.H.K. M.A., E.L. and F.F. All authors have read and agreed to the published version of the manuscript.

Funding: The authors extend their appreciation to the Deanship of Scientific Research at Majmaah University for funding this work under Project Number RGP-2019-32.

Acknowledgments: This work is partially supported by RTI2018-095499-B-C31, Funded by Ministerio de Ciencia, Innovación y Universidades, Gobierno de España (MCIU/AEI/FEDER,UE).

Conflicts of Interest: The authors declare no conflict of interest.

References

1. Alibakhshikenari, M.; Virdee, B.S.; See, C.H.; Abd-Alhameed, R.A.; Falcone, F.; Limiti, E. Surface wave reduction in antenna arrays using metasurface inclusion for MIMO and SAR systems. *Radio Sci.* **2019**, *54*, 1067–1075. [CrossRef]
2. Alibakhshikenari, M.; Virdee, B.S.; See, C.H.; Abd-Alhameed, R.A.; Falcone, F.; Limiti, E. High-isolation leaky-wave array antenna based on CRLH-metamaterial implemented on SIW with ±30° frequency beam-scanning capability at millimetre-waves. *Electronics* **2019**, *8*, 642. [CrossRef]
3. Alibakhshikenari, M.; Virdee, B.S.; Limiti, E. Triple-band planar dipole antenna for omnidirectional radiation. *Microw. Opt. Technol. Lett.* **2018**, *60*, 1048–1051. [CrossRef]
4. Alibakhshikenari, M.; Virdee, B.S.; See, C.H.; Abd-Alhameed, R.; Ali, A.; Falcone, F.; Limiti, E. Wideband printed monopole antenna for application in wireless communication systems. *IET Microw. Antennas Propag.* **2018**, *12*, 1222–1230. [CrossRef]
5. Alibakhshikenari, M.; Khalily, M.; Virdee, B.S.; See, C.H.; Abd-Alhameed, R.A.; Limiti, E. Mutual-coupling isolation using embedded metamaterial EM bandgap decoupling slab for densely packed array antennas. *IEEE Access* **2019**, *7*, 51827–51840. [CrossRef]
6. Jabire, A.H.; Zheng, H.X.; Abdu, A.; Song, Z. Characteristic mode analysis and design of wide band MIMO antenna consisting of metamaterial unit cell. *Electronics* **2019**, *8*, 68. [CrossRef]
7. Khan, M.S.; Shafique, M.F.; Capobianco, A.; Autizi, E.; Shoaib, I. Compact UWB-MIMO antenna array with a novel decoupling structure. In Proceedings of the 2013 10th International Bhurban Conference on Applied Sciences & Technology (IBCAST), Islamabad, Pakistan, 15–19 January 2013; pp. 347–350.
8. Irshad Khan, M.; Khattak, M.I.; Rahman, S.U.; Qazi, A.B.; Telba, A.A.; Sebak, A. Design and Investigation of Modern UWB-MIMO Antenna with Optimized Isolation. *Micromachines* **2020**, *11*, 432. [CrossRef]
9. Li, W.T.; Hei, Y.Q.; Subbaraman, H.; Shi, X.W.; Chen, R.T. Novel printed filtenna with dual notches and good out-of-band characteristics for UWB-MIMO applications. *IEEE Microw. Wirel. Compon. Lett.* **2016**, *26*, 765–767. [CrossRef]
10. Srivastava, K.; Kumar, A.; Kanaujia, B.K.; Dwari, S.; Kumar, S. A CPW-fed UWB MIMO antenna with integrated GSM band and dual band notches. *Int. J. RF Microw. Comput. Aided Eng.* **2019**, *29*, e21433. [CrossRef]
11. See, T.S.; Chen, Z.N. An ultrawideband diversity antenna. *IEEE Trans. Antennas Propag.* **2009**, *57*, 1597–1605. [CrossRef]
12. Najam, A.I.; Duroc, Y.; Tedjni, S. UWB-MIMO antenna with novel stub structure. *Prog. Electromagn. Res.* **2011**, *19*, 245–257. [CrossRef]

13. Iqbal, A.; Saraereh, O.A.; Ahmad, A.W.; Bashir, S. Mutual coupling reduction using F-shaped stubs in UWB-MIMO antenna. *IEEE Access* **2017**, *6*, 2755–2759. [CrossRef]

14. Saha, T.K.; Knaus, T.N.; Khosla, A.; Sekhar, P.K. A CPW-fed flexible UWB antenna for IoT applications. *Microsyst. Technol.* **2018**, *24*, 1–7. [CrossRef]

15. Jan, N.A.; Kiani, S.H.; Muhammad, F.; Sehrai, D.A.; Iqbal, A.; Tufail, M.; Kim, S. V-Shaped Monopole Antenna with Chichena Itzia Inspired Defected Ground Structure for UWB Applications. *CMC Comput. Mater. Contin.* **2020**, *65*, 19–32.

16. Sehrai, D.A.; Abdullah, M.; Altaf, A.; Kiani, S.H.; Muhammad, F.; Tufail, M.; Irfan, M.; Glowacz, A.; Rahman, S. A Novel High Gain Wideband MIMO Antenna for 5G Millimeter Wave Applications. *Electronics* **2020**, *9*, 1031. [CrossRef]

17. Simorangkir, R.B.; Kiourti, A.; Esselle, K.P. UWB wearable antenna with a full ground plane based on PDMS-embedded conductive fabric. *IEEE Antennas Wirel. Propag. Lett.* **2018**, *17*, 493–496. [CrossRef]

18. Yang, D.; Hu, J.; Liu, S. A low profile UWB antenna for WBAN applications. *IEEE Access* **2018**, *6*, 25214–25219. [CrossRef]

19. Li, M.; Luk, K.M. A differential-fed UWB antenna element with unidirectional radiation. *IEEE Trans. Antennas Propag.* **2016**, *64*, 3651–3656. [CrossRef]

20. Iqbal, A.; Saraereh, O.A.; Bouazizi, A.; Basir, A. Metamaterial-based highly isolated MIMO antenna for portable wireless applications. *Electronics* **2018**, *7*, 267. [CrossRef]

21. Iqbal, A.; Basir, A.; Smida, A.; Mallat, N.K.; Elfergani, I.; Rodriguez, J.; Kim, S. Electromagnetic bandgap backed millimeter-wave MIMO antenna for wearable applications. *IEEE Access* **2019**, *7*, 111135–111144. [CrossRef]

22. Elfergani, I.; Rodriguez, J.; Iqbal, A.; Sajedin, M.; Zebiri, C.; AbdAlhameed, R.A. Compact millimeter-wave MIMO antenna for 5G applications. In Proceedings of the 2020 14th European Conference on Antennas and Propagation (EuCAP), Copenhagen, Denmark, 15–20 March 2020; pp. 1–5.

23. Elfergani, I.; Iqbal, A.; Zebiri, C.; Basir, A.; Rodriguez, J.; Sajedin, M.; de Oliveira Pereira, A.; Mshwat, W.; Abd-Alhameed, R.; Ullah, S. Low-Profile and Closely Spaced Four-Element MIMO Antenna for Wireless Body Area Networks. *Electronics* **2020**, *9*, 258. [CrossRef]

24. Iqbal, A.; Smida, A.; Alazemi, A.J.; Waly, M.I.; Mallat, N.K.; Kim, S. Wideband Circularly Polarized MIMO Antenna for High Data Wearable Biotelemetric Devices. *IEEE Access* **2020**, *8*, 17935–17944. [CrossRef]

25. Kiani, S.H.; Mahmood, K.; Altaf, A.; Cole, A.J. Mutual coupling reduction of MIMO antenna for satellite services and radio altimeter applications. *Int. J. Adv. Comput. Sci. Appl.* **2018**, *9*, 23–26. [CrossRef]

26. Iqbal, A.; Altaf, A.; Abdullah, M.; Alibakhshikenari, M.; Limiti, E.; Kim, S. Modified U-Shaped Resonator as Decoupling Structure in MIMO Antenna. *Electronics* **2020**, *9*, 1321. [CrossRef]

27. Altaf, A.; Alsunaidi, M.A.; Arvas, E. A novel EBG structure to improve isolation in MIMO antenna. In Proceedings of the 2017 USNC-URSI Radio Science Meeting (Joint with AP-S Symposium), San Diego, CA, USA, 9–14 July 2017; pp. 105–106.

28. Jaglan, N.; Gupta, S.D.; Thakur, E.; Kumar, D.; Kanaujia, B.K.; Srivastava, S. Triple band notched mushroom and uniplanar EBG structures based UWB MIMO/Diversity antenna with enhanced wide band isolation. *AEU-Int. J. Electron. Commun.* **2018**, *90*, 36–44. [CrossRef]

29. Gorai, A.; Ghatak, R. Utilization of Shorted Fractal Resonator topology for high isolation and ELC resonator for band suppression in compact MIMO UWB antenna. *AEU-Int. J. Electron. Commun.* **2020**, *113*, 152978. [CrossRef]

30. Li, J.F.; Chu, Q.X.; Li, Z.H.; Xia, X.X. Compact dual band-notched UWB MIMO antenna with high isolation. *IEEE Trans. Antennas Propag.* **2013**, *61*, 4759–4766. [CrossRef]

31. Zhang, S.; Lau, B.K.; Sunesson, A.; He, S. Closely-packed UWB MIMO/diversity antenna with different patterns and polarizations for USB dongle applications. *IEEE Trans. Antennas Propag.* **2012**, *60*, 4372–4380. [CrossRef]

32. Cruz, J.D.N.; Serres, A.J.R.; de Oliveira, A.C.; Xavier, G.V.R.; de Albuquerque, C.C.R.; da Costa, E.G.; Freire, R.C.S. Bio-inspired Printed Monopole Antenna Applied to Partial Discharge Detection. *Sensors* **2019**, *19*, 628. [CrossRef]

33. Lin, G.S.; Sung, C.H.; Chen, J.L.; Chen, L.S.; Houng, M.P. Isolation improvement in UWB MIMO antenna system using carbon black film. *IEEE Antennas Wirel. Propag. Lett.* **2016**, *16*, 222–225. [CrossRef]

34. Iqbal, A.; Selmi, M.A.; Abdulrazak, L.F.; Saraereh, O.A.; Mallat, N.K.; Smida, A. A Compact Substrate Integrated Waveguide Cavity-Backed Self-Triplexing Antenna. *IEEE Trans. Circuits Syst. II Express Briefs* **2020**. [CrossRef]
35. Iqbal, A.; Alazemi, A.J.; Mallat, N.K. Slot-DRA-based independent dual-band hybrid antenna for wearable biomedical devices. *IEEE Access* **2019**, *7*, 184029–184037. [CrossRef]
36. Iqbal, A.; Bouazizi, A.; Kundu, S.; Elfergani, I.; Rodriguez, J. Dielectric resonator antenna with top loaded parasitic strip elements for dual-band operation. *Microw. Opt. Technol. Lett.* **2019**, *61*, 2134–2140. [CrossRef]

Article

Compact Rectifier Circuit Design for Harvesting GSM/900 Ambient Energy

Surajo Muhammad [1], Jun Jiat Tiang [1], Sew Kin Wong [1], Amjad Iqbal [1,*], Mohammad Alibakhshikenari [2,*] and Ernesto Limiti [2]

[1] Centre For Wireless Technology (CWT), Faculty of Engineering, Multimedia University, Cyberjaya 63100, Malaysia; doguwa_2002@yahoo.com (S.M.); jjtiang@mmu.edu.my (J.J.T.); skwong@mmu.edu.my (S.K.W.)

[2] Electronic Engineering Department, University of Rome "Tor Vergata", Via del Politecnico 1, 00133 Rome, Italy; limiti@ing.uniroma2.it

* Correspondence: amjad730@gmail.com (A.I.); Alibakhshikenari@ing.uniroma2.it (M.A.); Tel.: +60-1128-784-475 (A.I.)

Received: 6 September 2020; Accepted: 29 September 2020; Published: 1 October 2020

Abstract: In this paper, a compact rectifier, capable of harvesting ambient radio frequency (RF) power is proposed. The total size of the rectifier is 45.4 mm $\times$ 7.8 mm $\times$ 1.6 mm, designed on FR-4 substrate using a single-stage voltage multiplier at 900 MHz. GSM/900 is among the favorable RF Energy Harvesting (RFEH) energy sources that span over a wide range with minimal path loss and high input power. The proposed RFEH rectifier achieves measured and simulated RF-to-dc (RF to direct current) power conversion efficiency (PCE) of 43.6% and 44.3% for 0 dBm input power, respectively. Additionally, the rectifier attained 3.1 V DC output voltage across 2 kΩ load terminal for 14 dBm and is capable of sensing low input power at -20 dBm. The work presents a compact rectifier to harvest RF energy at 900 MHz, making it a good candidate for low powered wireless communication systems as compares to the other state of the art rectifier.

Keywords: RF energy harvesting; impedance matching network (IMN); power conversion efficiency; rectifier; IoT

---

1. Introduction

The ubiquitous nature of wireless communication technology has over the years attract and gained much attention and contribution from researchers globally [1,2]. Additionally, GSM900/1800, 3G, 4G, Wi-Fi, and the evolving auspicious 5G are among a few dominant wireless technology generation operating daily to cater for human needs and demand [3,4]. Recent development in wireless communication technology gave birth to new and evolving technological innovations. The increasing demand of such invention led to one or more application driver in the areas of IoT, industrial IoT, autonomous driving, smart farming, smart cities, and many more [5,6]. Uninterrupted power supply over remote places limits the potential of wire-based ultra-low powered IoT devices and sensor nodes. which directly restricts their capabilities for security surveillances, smart farming, and other related applications besides a copper loss [5–7]. One of the main challenges with the evolving ultra-low powered IoT devices is energy storage limitation. These, in turn, increase the need for RFEH systems. To over come the major challenges with traditional battery-based systems to charge/maintain/replace a battery. RFEH is believed to be a promising technology through direct power or battery recharging from abundant electromagnetic (EM) energy radiating in the environment. The technology can be deployed in low-powered devices for security surveillance, smart farming and many more [5–8]. Initially, the concept of energy harvesting started to evolve in the 1990s along with Tesla finding's on electrical energy conversion from an EM wave [5]. RF power system typically comprises an antenna, IMN, a rectifying diode, a storage element (DC pass filter), and a terminal load. A combination of five circuit elements gives rise to rectenna. A rectenna without the antenna component is categorized as a rectifying circuits. Figure 1 provides a block diagram containing the key components of rectenna. The matching network (MN) and the rectifier circuit play an important role in the RFEH system. MN is aim at ensuring optimum energy transfer between the receiving antenna and the load, and the latter transform the scavenged RF EM signals into a usable DC output voltage. RFEH module can be deployed in a sensitive and critical environment such as autonomous driving, security surveillance, and healthcare systems, for their capability to operate under both environmental and nonenvironmental condition unlike the other forms of energy harvesting system [5]. The operation of rectifying circuit and RFEH system are generally assessed through (PCE). Several RFEH circuits design have been studied and proposed to achieve significant PCE through the use of wide-band, multi-band, dual and single band rectifiers. Among the designs, single-band rectenna used to be more efficient as compared to their wide or broad band counter parts [5,8]. The broadband rectifiers are generally associated with much more circuits complexity, which reduces the overall circuit performance because of parasitics capacitance at the junction of the diode and other lumped elements [8,9]. The authors of [10,11] designed a rectenna with maximum PCE at 37.8% and 28% using −1.5 dBm and −15 dBm input power. To recycle ambient RF signal, a rectenna with 20% efficiency is described in [12] using 62 μW/cm^2 (13.27 dBm) input power with a total geometry of 342.55 cm^2. Several research projects have been developed to accomplish efficient rectenna at relatively low input power [13,14]. An RF rectifier is proposed in [15], the design can achieve up to 33% efficiency at −10 dBm input power at the cost of complex circuit with large electrical length. A maximum PCE of 47.8% and 38% is highlighted in [16,17], using an input power between (5–14 dBm), respectively. The authors of [13] also worked on rectifier circuit and realized 50% maximum PCE using 21 dBm solar panel. A rectenna that achieved up to 24% PCE is proposed in [14], using −20 dBm input power with help of an antenna array and rectifier with a total geometry of 32.4 cm^2.

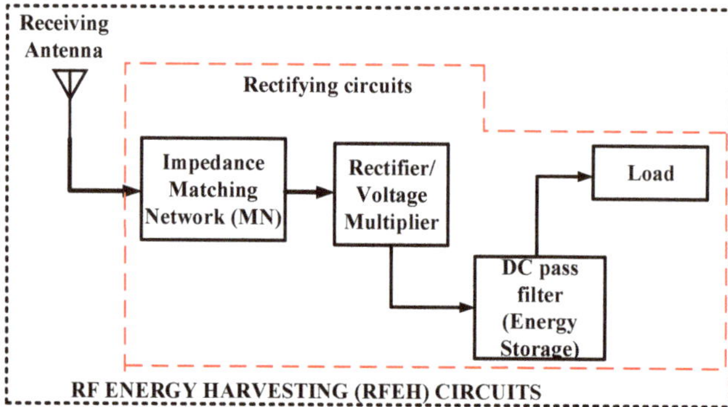

Figure 1. Block diagram of RF energy harvesting circuits.

Many foregoing types of research in the RFEH system focused on a much higher frequency between GSM/1800 to WLAN/5200 to achieve compactness. The operational frequency contributed significantly to achieving a compact RFEH circuit [7–18]. The RF spectral survey from the open literature and the one carried out in this work, using (Aim TTi PSA6005 6 GHz RF spectrum analyzer), come out with a notable result surrounding the lower GSM/900 band towards generating higher available ambient RF power [14,15]. This paper, proposed a compact 3.51 cm^2 rectifier at 900 MHz using simple L-section MN. Conversely, the designed is a good candidate for harvesting RF energy in ambient environment with capability of sensing RF input power below −20 dBm integrable with low powered IoT devices. In the remaining sections of this work, Section 2 presents a rectifier circuit design and analysis, Section 3 highlights the result of a finding of the proposed rectifier and Section 4 provides the concluding remarks. The following outlines the main contribution of this work:

- To the best of our knowledge, this is the compact RFEH rectifier as compares to state of the art rectifiers at 900 MHz.
- This research targets an energy harvesting rectifier that works over GSM/900 to cover long and remote places suitable for low powered IoT devices and sensor nodes.
- The proposed design can receive an RF input signal at less than -20 dBm and also provides detailed circuit and performance analysis.

2. Rectifier Circuit Design and Analysis

The voltage appears at the end of an RFEH antenna is considerably of small quantity and as well analogous to sinusoidal signal.A conditioning circuit across the load terminal is an important figure to improve the output DC signals of the RFEH module. Diode(s), MN, and load terminal are primary components to the rectifying circuits. The diode(s) is playing a vital role in converting the accessible ambient RF power into a usable DC supply. The MN minimizes the reflection losses for RF input power received by the antenna. The low powered RFEH MN need to be simple in design with minimum losses.

2.1. Transmission Line and Input Impedance Matching Network

A 50 Ω microstrip transmission line width and length to match the antenna with MN and voltage multiplier was first to design using FR-4 substrate ($\epsilon_r = 5.4$, h = 1.6 mm, and tan$\delta = 0.02$). The substrate is chosen for its low cost, light weight, and ease of fabrication [7]. ADS LineCal operator is applied to compute and optimized the proposed width and length of the line at 2.7 mm and 30 mm, respectively. MN is significant part of the rectifier circuits. The design process of MN is a challenging task because the rectifier input impedance (Z_{in}) is controlled by: operating frequency (f_c), input power (P_{in}), diode inconstant internal resistance (R_d) and terminal load (R_L) [19].

Figure 2 shows L-section MN to match the impedances between the source and the load. The proposed model reduces circuit complexity using two simple inductive (L_m) and capacitive (C_m) reactive elements. The inductor (L_m) plays a vital role towards enhancing the input signal getting into the rectifier and the capacitor (C_m) enhances the rectifier matching [19]. The MN parameters can be computed by negating source and load imaginary part and equating their real part at the design frequency as illustrated in Equations (1)–(5). The equivalent power source from the antenna (P_s) is analogous to the antenna available power (P_{av}) passing through the transmission line and the MN. Other related reference power notation in the circuit model includes rectifier input power (P_{in}) and the output power consumed by the load terminal (P_L).

Equating the real part of the antenna with the rectifier input impedance we get:

$$R_{ant} = R_{in} \cdot \left(\frac{1}{1+Q^2} \right) \tag{1}$$

where R_{in} = rectifier input impedance, R_{ant} = source impedance, and the quality factor Q is given by:

$$Q = \sqrt{\frac{R_{in}}{R_{ant}} - 1} \tag{2}$$

To determine the reactance parameters, the quality factor Q can be express as the ratio of imaginary (reactive) part of the impedance to the respective real (resistive) components as:

$$Q = \frac{Im(Z)}{Re(Z)} = \frac{Im\left(\dfrac{1}{\frac{1}{R_{in}} + j\omega_0 C_{in} + \frac{1}{j\omega_0 L_m}} \right)}{Re\left(\dfrac{1}{\frac{1}{R_{in}} + j\omega_0 C_{in} + \frac{1}{j\omega_0 L_m}} \right)} \tag{3}$$

$$= \frac{R_{in}}{\omega_o L_m} - \omega_o C_{in} R_{in} \tag{3}$$

where C_{in} = reactive components of the load impedance operating at frequency ω_o, using Equation (3) the inductance L_m of the network is given as:

$$L_m = \frac{R_{in}}{\omega_o (Q + \omega_o C_{in} R_{in})} \tag{4}$$

L-section capacitance C_m can be computed by setting the imaginary part of Equation (3) to zero as:

$$C_m = \frac{R_{in}}{L_m(R_{in} - R_{ant})} \cdot \frac{1}{\left(\omega_o^2 - \frac{1}{L_m C_{in}}\right)} \tag{5}$$

A two-element MN (L1 and C1) is designed at 900 MHz using a 50 Ω transmission line as (R_{ant}), terminated with a 2 kΩ load. The numerical values of $L1$ and $C1$ were computed at 55 nH and 0.5 pF, respectively. The parameters were transfer into ADS simulation software using ideal component palette. Because of the effects of the transmission line in the circuits, the lumped elements need to be tune and optimize. ($L1$) and ($C1$) were compensated with -10 nH and $+0.5$ pF, and the rectifier is matched at 45 nH and 1 pF, respectively. The circuit is further tuned by replacing the ideal elements with *muRata* equivalent surface mount device (SMD) in the ADS library at 27 nH and 2 pF.

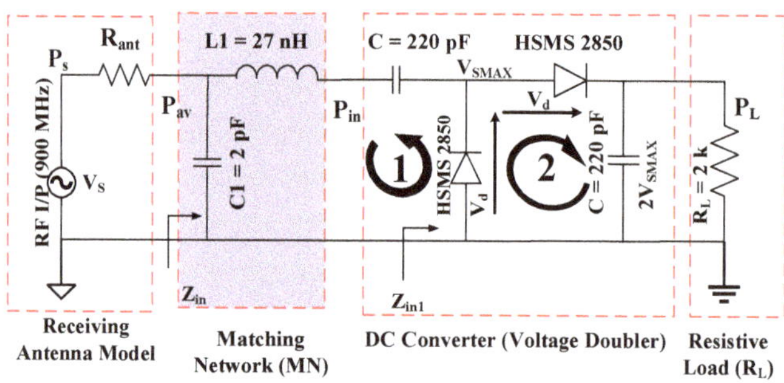

Figure 2. Detailed circuit configuration of the proposed rectifier.

2.2. Rectifier Design

Rectifying circuits are of different configuration layout. The circuits can be achieved as a single series or shunt rectifier diode, voltage doubler rectifiers, up to the level of multistage and cascaded rectifiers. The authors of [9] highlights an extensive study on various types of rectifier circuits topology. Rectifier selection to operate at high frequency need to have: reliable power response, and the ability to manage power with minimal dissipation [7] for efficient and effective power harvesting. A single-stage voltage multiplier configuration is a suitable candidate to be deployed in RFEH circuits as compared to a single series/shunt half-wave configurations. An RF rectifier circuit with a single-stage voltage multiplier is more efficient with higher output power compares with single series/shunt half-wave rectifiers. Single-stage voltage multiplier rectifiers also offer fewer losses in contrast to multistage rectifiers for a low-powered RF harvester [9]. The proposed rectifier is built based on a voltage multiplier topology with a pair of HSMS-2850 diode (SOT-323 layout), as shown in Figure 2. The diode is a suitable candidate for low power RF application with a small junction capacitance of 0.18 pF and a minimum detectable voltage of 150 mV at 0.1 mA [5]. The rectifier junction capacitance (C) is controlled by the operating frequency f_c and voltage as express in Equation (6).

$$C = I \cdot \frac{dt}{dV} \tag{6}$$

where C is the capacitance of the capacitor with current I passing through junction diode at the rate of non linear voltage "V".

The antenna then receives and transmits the reactive AC power into the rectifier as express by the equivalent wave Equation (7) [9]. During the negative half-wave AC circle diode $D1$, become forward bias and charges (C) in label (1) to (V_{SMAX}). At the positive half-wave AC cycle, diode $D1$ is reverse bias blocking the DC voltage in label (1) from discharging, and diode $D2$ in label (2) change into forward bias. The reserve (V_{SMAX}) in label (1) and diode $D2$ synchronously sum up the output DC voltage to ($2V_{SMAX}$) onto the capacitor (C) in label (2), which discharges through the load terminal (R_L) as describe in Equations (8) and (9).

$$V_S = V_{SMAX} sin\omega t \tag{7}$$

where V_{SMAX} is the amplitude of the RF input signal received at the output of the MN operating at frequency ω [20].

$$-V_{SMAX} - V_{C1} + V_{C2} + 2V_d = 0 \tag{8}$$

where V_d is the forward bias voltage of the rectifying diode, and

$$V_{C2} = V_{SMAX} + V_{C1} - 2V_d, and$$

$$V_{C1} = V_{SMAX}$$

$$V_{C2} = 2V_{SMAX} - 2V_d \tag{9}$$

3. Results and Discussion

The propose rectifier is fabricated on a low-cost commercial FR-4 board with a total dimension 45.4 mm $\times$ 7.8 mm $\times$ 1.6 mm connected via 50 Ω SMA probe. A pair of small crocodile clip is used to connect the rectifier across the load terminal to measure the output parameters as seen in Figure 3b. Figure 3a also provides layout model of the rectifier prototype.

Equation (10) provides a wheeler's transmission line width equation that portrays the effects of (W) on the rectifier matching [21]. Decreasing (W) below the designed value causes a direct rectifier mismatch at the proposed frequency. Increasing (W) above the designed value shifted the rectifier operating frequency by 20 MHz in addition to impedance mismatch. The transmission line length is computed through a guided wave at $\lambda_g/4 = 42.5$ mm and is optimized at 30 mm.

$$Z_0 = \frac{120\pi}{\sqrt{\epsilon_e}} \left[\frac{W}{h} + 1.393 + 0.667 \ln(\frac{W}{h} + 1.444) \right]^{-1} \tag{10}$$

where Z_o is the characteristic impedance of the line, W = width of transmission line, h = height of the substrate, and ϵ_e gives the effective dielectric constant of the substrate.

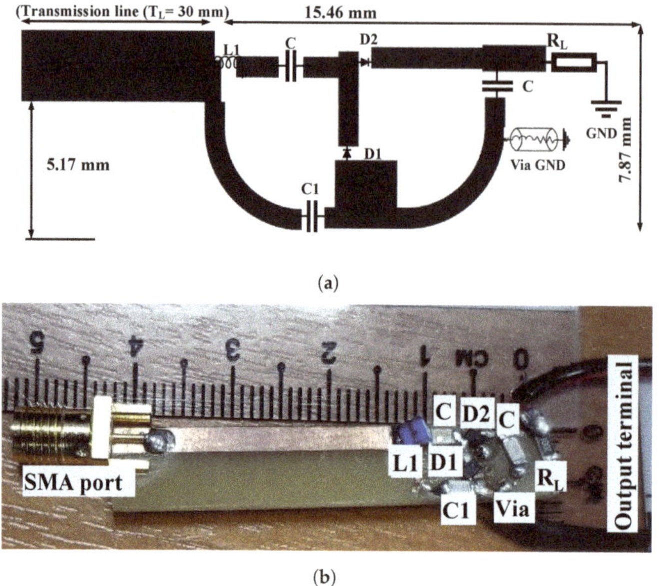

(a)

(b)

Figure 3. Rectifier proposed topology: (**a**) EM model layout, (**b**) fabricated prototype.

To evaluate the rectifier performance at the design frequency, L-section MN undergo a parametric study. Figure 4a depicts the simulated reflection coefficient versus frequency at different values of the inductor (*L1*). The design frequency shifted to about 1.1 GHz for an inductance value from 27 nH to 18 nH. Following the inductive reactance properties, frequency increases the current flowing through the inductor, which lowers the inductance value in the circuits [3–5,13]. Figure 4b also shows the effect of capacitance (*C1*) in the MN, varying the capacitance from 2 pF to 10 pF deteriorates the rectifier matching. Hence, lowering or increasing the capacitance value leads to rectifier mismatched which can render the rectifying circuit inefficient [9,20].

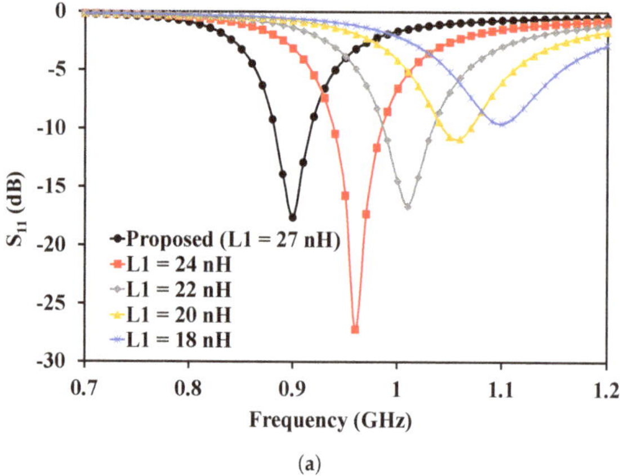

(a)

Figure 4. *Cont.*

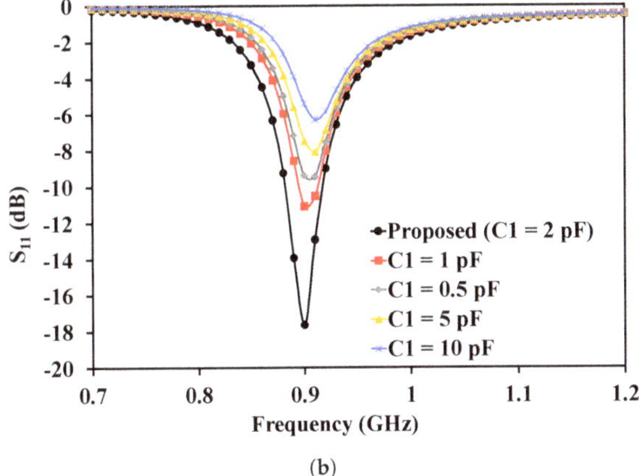

Figure 4. Simulated reflection coefficient of the rectifier for: (**a**) Inductance (*L*1) (**b**) Capacitance (*C*1).

To analyze the PCE of the proposed rectifier, Figure 5a presents the result of the load terminal (R_L) sweep at various input power levels (from −15 dBm to 10 dBm in step of 5 dBm). The result shows that the efficiency increases with an increase of input power across the load which agreed well with Equation (7) [5–22]. The rectifier achieves a maximum PCE along the 2–4 kΩ load terminal in both cases. The design shows a low PCE for −15 dBm with an efficiency greater than 20% for a load above 2.5 kΩ. A 60% high efficiency is also observed around the 2–4 kΩ load for +10 dBm input power.

Figure 5b presents the simulated and measured reflection coefficient (S_{11}) of the proposed rectifier at 900 MHz. The design achieved a simulated and measured −10 dB bandwidth of 45 MHz (880–925 MHz) and 46 MHz (870–916 MHz) respectively. A vector network analyzer (E5062A) from Agilent Technologies is used to measure the (S_{11}) of the rectifier for −20 dBm input power level.

Equation (12) provides a relation that describes the rectifier ability in converting the received RF signal in to a usable output DC signal known as (PCE) in terms of output and input power. To determine the simulated output DC voltage and efficiency over a range of frequency, input power and a terminal load. A large signal s-parameter (LSSP) and harmonic balance (HB) simulation modules's that coexist in ADS is deployed into the design owing to the non-linear characteristics of the rectifying diode [5–11]. A 12 GHz APSIN12G signal generator sweep from −30 dBm to 10 dBm input power is used to measure the rectifier output DC voltage across the 2 kΩ load terminal. The measured result is in good agreement with simulation, as depicted in Figure 6a. The rectifier can receive an RF input signal at −30 dBm with the capability of harvesting 3.1 V for both simulated and measured results at 14 dBm and 16 dBm, respectively. Equation (11) provides the rectifier output DC voltage from Equation (9) [9,23].

$$V_{DC} = V_{C2} \tag{11}$$

$$\eta_{PCE}\% = \frac{P_{DC}}{P_{in}} \times 100\% = (\frac{V_{DC}^2}{R_L}) \times (\frac{1}{P_{in}}) \times 100\% \tag{12}$$

where η_{PCE} is the power conversion efficiency, P_{DC} is DC output power across the load R_L and V_{DC} gives the output DC voltage.

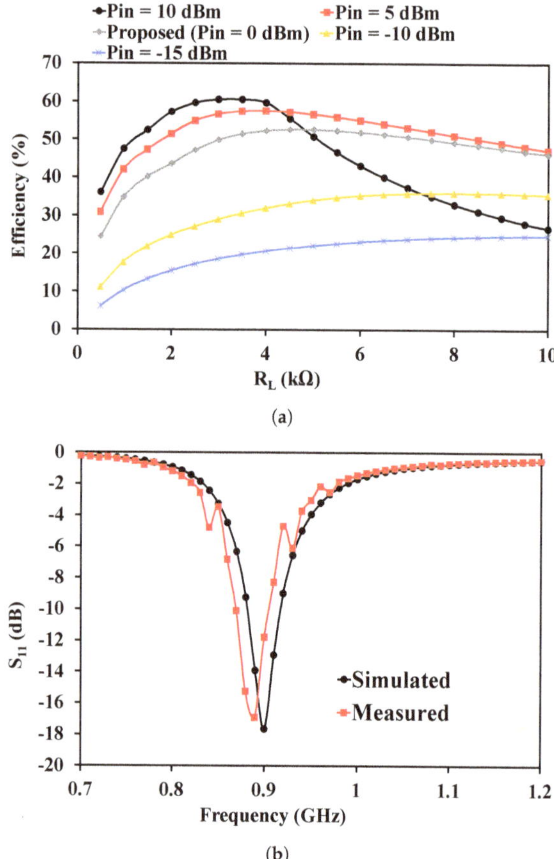

Figure 5. (**a**) Simulated efficiency sweep against load terminal (R_L) at various input power (P_{in}). (**b**) Simulated and measured reflection coefficient of the proposed rectifier for -20 dBm input power.

Figure 6b presents the simulated and measured PCE across the 2 kΩ load terminal, and the rectifier achieves a PCE of 44.3% and 43.6% for 0 dBm input power, respectively. The rectifier experiences a drop in measured efficiency because of nonlinear characteristics of the diode as reported in the datasheet, in connection with other losses from the circuit board (Such as SMA connector probe, interconnecting transmission lines, soldering, and chip-component tolerant losses) [3].

Table 1 provides a comparison table of the proposed design with other related works in the literature. The authors of [10,11] presents an energy harvester with about times larger in electrical length than the proposed design. The authors of [13] did not record the amount of input power at which the rectifier achieves its maximum PCE, but uses a solar antenna [24] with 21 dBm capacity and a much larger rectifier. The authors of [14] also comes out with a harvesting circuit without describing the MN used besides rectifier large electrical length. A much lower PCE at high input power is reported in [16,17]. The proposed design in the open literature target high-frequency band rectifier starting from GSM/1800—WLAN/5200 which contributes to achieving rectifier compactness. The proposed design in this work offers a compact size rectifier capable of harvesting RF signal over a long-range cover by GSM/900.

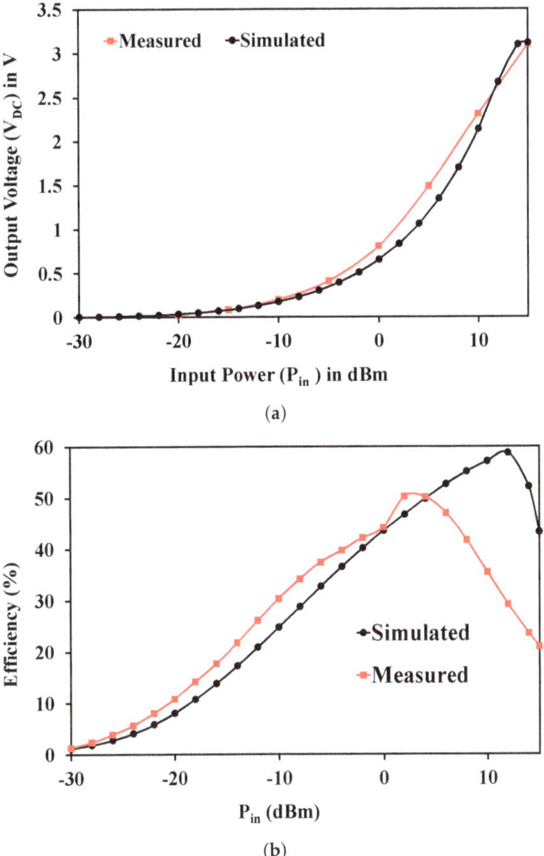

(a)

(b)

Figure 6. Simulated and measured: (**a**) Output DC voltage sweep against the input power (**b**) PCE sweep over an input power P_{in} at 900 MHz.

Table 1. Comparison of the proposed rectifier and related.

Ref	Electrical Size (λ_g)	η_{PCE} (%) @P_{in} (dBm)	Frequency (GHz)	Diode
[10]	$1.03\lambda_g \times 1.03\lambda_g$	37.8%@-1.5	2.45	HSMS286C
[11]	$0.76\lambda_g \times 0.84\lambda_g$	28%@-15	2.45	HSMS2850
[13]	$3.86\lambda_g \times 4.89\lambda_g$	50%@NA	1.8/2.1	SMS7630
[14]	$0.88\lambda_g \times 0.57\lambda_g$	24%@-20	1.8/2.5	HSMS2850
[16]	$0.34\lambda_g \times 0.21\lambda_g$	(47.8, 37, 46.7, 42)% @(14, 12, 13, 12)	0.87/1.27/ 2.02/2.38	HSMS2860
[17]	$0.38\lambda_g \times 0.22\lambda_g$	(38, 37, 18, 12.5)% @5	1.3/1.7/ 2.4/3.6	SMS7630
This work	$0.31\,\lambda_g \times 0.05\,\lambda_g$	50.2%@2	0.90	HSMS2850

4. Conclusions

This paper, proposes the design and construction of a compact rectifier at a 900 MHz GSM band, using HSMS-2850 Schottky diode. A transmission line and an L-section impedance matching network are designed from the closed-form equations, to match the input impedance of the rectifier with 50 Ω. The circuit undergoes parametric tuning from an ideal to model component to achieve better performance and minimizes the effects of the transmission line in the network. The fabricated prototype can receive an ambient RF signal below -20 dBm and achieved 50.2% and 49.1% maximum PCE for both simulated and measured results at 14 dBm and 2 dBm, respectively. Similarly, the circuit reaches an output DC voltage of 3.1 V for simulated and measured results at 14 dBm and 16 dBm, respectively using a 2 kΩ load terminal. The proposed rectifier occupies a total of 351 mm^2 ($0.31\lambda_g \times 0.05\lambda_g$) on the PCB circuit board, offering efficiency and compactness improvement as compares to other recent designs making it suitable for various compact energy harvesting applications.

Author Contributions: Conceptualization, S.M.; methodology, S.M. and M.A.; software, S.M.; validation, S.M., A.I., M.A. and E.L.; formal analysis, S.M.; investigation, S.M.; resources, J.J.T. and S.K.W.; data curation, S.M.; writing—original draft preparation, S.M.; writing—review and editing, S.M. and A.I.; visualization, S.M., M.A. and E.L.; supervision, J.J.T. and S.K.W.; project administration, J.J.T. and S.K.W.; funding acquisition, J.J.T. and S.K.W. All authors have read and agreed to the published version of the manuscript.

Funding: This work was supported by TM R&D Malaysia under project number MMUE/190001.

Conflicts of Interest: The authors declare no conflict of interest.

References

1. Chen, Y.S.; You, J.W. A scalable and multidirectional rectenna system for RF energy harvesting. *IEEE Trans. Compon. Packag. Manuf. Technol.* **2018**, *8*, 2060–2072. [CrossRef]
2. Pardue, C.A.; Bellaredj, M.L.F.; Davis, A.K.; Swaminathan, M.; Kohl, P.; Fujii, T.; Nakazawa, S. Design and characterization of inductors for self-powered IoT edge devices. *IEEE Trans. Compon. Packag. Manuf. Technol.* **2018**, *8*, 1263–1271. [CrossRef]
3. Mansour, M.M.; Kanaya, H. Compact and broadband RF rectifier with 1.5 octave bandwidth based on a simple pair of L-Section matching network. *IEEE Microw. Wirel. Compon. Lett.* **2018**, *28*, 335–337. [CrossRef]
4. Pardue, C.A.; Bellaredj, M.L.F.; Torun, H.M.; Swaminathan, M.; Kohl, P.; Davis, A.K. RF Wireless Power Transfer Using Integrated Inductor. *IEEE Trans. Compon. Packag. Manuf. Technol.* **2018**, *9*, 913–920. [CrossRef]
5. Awais, Q.; Jin, Y.; Chattha, H.T.; Jamil, M.; Qiang, H.; Khawaja, B.A. A compact rectenna system with high conversion efficiency for wireless energy harvesting. *IEEE Access* **2018**, *6*, 35857–35866. [CrossRef]
6. Wong, S.W.; Sun, G.H.; Zhu, L.; Chen, Z.N.; Chu, Q.X. Integration of Wireless Coil and Bluetooth Antenna for High Charging and Radiation Efficiencies. *IEEE Trans. Compon. Packag. Manuf. Technol.* **2018**, *8*, 1292–1299. [CrossRef]
7. Mansour, M.M.; Kanaya, H. Novel L-Slot Matching Circuit Integrated with Circularly Polarized Rectenna for Wireless Energy Harvesting. *Electronics* **2019**, *8*, 651. [CrossRef]
8. Liu, J.; Zhang, X.Y. Compact triple-band rectifier for ambient RF energy harvesting application. *IEEE Access* **2018**, *6*, 19018–19024. [CrossRef]
9. Song, C.; Huang, Y.; Carter, P.; Zhou, J.; Yuan, S.; Xu, Q.; Kod, M. A novel six-band dual CP rectenna using improved impedance matching technique for ambient RF energy harvesting. *IEEE Trans. Antennas Propag.* **2016**, *64*, 3160–3171. [CrossRef]
10. Huang, F.J.; Yo, T.C.; Lee, C.M.; Luo, C.H. Design of circular polarization antenna with harmonic suppression for rectenna application. *IEEE Antennas Wirel. Propag. Lett.* **2012**, *11*, 592–595. [CrossRef]
11. Palazzi, V.; Kalialakis, C.; Alimenti, F.; Mezzanotte, P.; Roselli, L.; Collado, A.; Georgiadis, A. Design of a ultra-compact low-power rectenna in paper substrate for energy harvesting in the Wi-Fi band. In Proceedings of the 2016 IEEE Wireless Power Transfer Conference (WPTC), Aveiro, Portugal, 5–6 May 2016; pp. 1–4.
12. Falkenstein, E.; Roberg, M.; Popovic, Z. Low-power wireless power delivery. *IEEE Trans. Microw. Theory Tech.* **2012**, *60*, 2277–2286. [CrossRef]
13. Niotaki, K.; Georgiadis, A.; Collado, A.; Vardakas, J.S. Dual-band resistance compression networks for improved rectifier performance. *IEEE Trans. Microw. Theory Tech.* **2014**, *62*, 3512–3521. [CrossRef]

14. Adam, I.; Yasin, M.N.M.; Rahim, H.A.; Soh, P.J.; Abdulmalek, M.F. A compact dual-band rectenna for ambient RF energy harvesting. *Microw. Opt. Technol. Lett.* **2018**, *60*, 2740–2748. [CrossRef]

15. Masotti, D.; Costanzo, A.; Francia, P.; Filippi, M.; Romani, A. A load-modulated rectifier for RF micropower harvesting with start-up strategies. *IEEE Trans. Microw. Theory Tech.* **2014**, *62*, 994–1004. [CrossRef]

16. Lu, J.J.; Yang, X.X.; Mei, H.; Tan, C. A four-band rectifier with adaptive power for electromagnetic energy harvesting. *IEEE Microw. Wirel. Compon. Lett.* **2016**, *26*, 819–821. [CrossRef]

17. Hsu, C.Y.; Lin, S.C.; Tsai, Z.M. Quadband rectifier using resonant matching networks for enhanced harvesting capability. *IEEE Microw. Wirel. Compon. Lett.* **2017**, *27*, 669–671. [CrossRef]

18. Vuong, T.P.; Verdier, J.; Allard, B.; Benech, P. Design and Measurement of 3D Flexible Antenna Diversity for Ambient RF Energy Scavenging in Indoor Scenarios. *IEEE Access* **2019**, *7*, 17033–17044.

19. Papotto, G.; Carrara, F.; Palmisano, G. A 90-nm CMOS threshold-compensated RF energy harvester. *IEEE J. Solid-State Circuits* **2011**, *46*, 1985–1997. [CrossRef]

20. Singh, N.; Kanaujia, B.; Beg, M.; Mainuddin; Kumar, S.; Choi, H.C.; Kim, K.W. Low Profile Multiband Rectenna for Efficient Energy Harvesting at Microwave Frequencies. *Int. J. Electron.* **2019**, *106*, 2057–2071. [CrossRef]

21. Balanis, C.A. *Antenna Theory: Analysis and Design*; John Wiley & Sons: Hoboken, NJ, USA, 2016.

22. Hagerty, J.A.; Helmbrecht, F.B.; McCalpin, W.H.; Zane, R.; Popovic, Z.B. Recycling ambient microwave energy with broad-band rectenna arrays. *IEEE Trans. Microw. Theory Tech.* **2004**, *52*, 1014–1024. [CrossRef]

23. Tissier, J.; Latrach, M. A 900/1800 MHz dual-band high-efficiency rectenna. *Microw. Opt. Technol. Lett.* **2019**, *61*, 1278–1283. [CrossRef]

24. Niotaki, K.; Collado, A.; Georgiadis, A.; Kim, S.; Tentzeris, M.M. Solar/electromagnetic energy harvesting and wireless power transmission. *Proc. IEEE* **2014**, *102*, 1712–1722. [CrossRef]

 *electronics*

Article

Infinity Shell Shaped MIMO Antenna Array for mm-Wave 5G Applications

Mian Muhammad Kamal [1,*], Shouyi Yang [1], Xin-cheng Ren [2], Ahsan Altaf [3], Saad Hassan Kiani [4], Muhammad Rizwan Anjum [5], Amjad Iqbal [6], Muhammad Asif [4] and Sohail Imran Saeed [7,*]

1 School of Information Engineering, Zhengzhou University, Zhengzhou 450001, China; iesyyang@zzu.edu.cn
2 School of Physics and Electronic Information, Yanan University, Shaanxi 761000, China; xchren@yau.edu.cn
3 Electrical Engineering Department, Istanbul Medipol University, 34083 Istanbul, Turkey; aaltaf@st.medipol.edu.tr
4 Electrical Engineering Department, City University of Science and Information Technology, Peshawar 25000, Pakistan; saad.kiani@cusit.edu.pk (S.H.K.); masifee@cusit.edu.pk (M.A.)
5 Department of Electronic Engineering, The Islamia University of Bahawalpur, Bahawalpur 63100, Pakistan; engr.rizwan@iub.edu.pk
6 Faculty of Engineering, Multimedia University, Cyberjaya 63100, Malaysia; aiqbal@ieee.org
7 Electrical Engineering Department, Iqra National University, Peshawar 25000, Pakistan
* Correspondence: mmkamal@gs.zzu.edu.cn (M.M.K.); sohail.imran@inu.edu.pk (S.I.S.)

Abstract: In this paper, a novel single layer Multiple Input–Multiple Output (MIMO) antenna for Fifth-Generation (5G) 28 GHz frequency band applications is proposed and investigated. The proposed MIMO antenna operates in the Ka-band, which is the most desirable frequency band for 5G mm-wave communication. The dielectric material is a Rogers-5880 with a relative permittivity, thickness and loss tangent of 2.2, 0.787 mm and 0.0009, respectively, in the proposed antenna design. The proposed MIMO configuration antenna element consists of triplet circular shaped rings surrounded by an infinity-shaped shell. The simulated gain achieved by the proposed design is 6.1 dBi, while the measured gain is 5.5 dBi. Furthermore, the measured and simulated antenna efficiency is 90% and 92%, respectively. One of the MIMO performance metrics—i.e., the Envelope Correlation Coefficient (ECC)—is also analyzed and found to be less than 0.16 for the entire operating bandwidth. The proposed MIMO design operates efficiently with a low ECC, better efficiency and a satisfactory gain, showing that the proposed design is a potential candidate for mm-wave communication.

Keywords: 5G antenna; shell; gain; 5G; devices; directivity; bandwidth

Citation: Kamal, M.M.; Yang, S.; Ren, X.-c.; Altaf, A.; Kiani, S.H.; Anjum, M.R.; Iqbal, A.; Asif, M.; Saeed, S.I. Infinity Shell Shaped MIMO Antenna Array for mm-Wave 5G Applications. *Electronics* **2021**, *10*, 165. https://doi.org/10.3390/electronics10020165

Received: 23 November 2020
Accepted: 6 January 2021
Published: 13 January 2021

Publisher's Note: MDPI stays neutral with regard to jurisdictional claims in published maps and institutional affiliations.

1. Introduction

Fifth Generation (5G) communication has now become the most discussed technology. The reason for the great attraction of 5G is the continuously progressing demand for higher data rates and bandwidth, which is not possible to fulfill with the current standards below the 5G technology [1,2]. The limited bandwidth availability in the microwave region and highly busy portion of the spectrum have forced a move towards a new spectrum range which can offer data rates in the range of multi-gigabits per second (Gbps) [3]. 5G is currently divided into sub-6 GHz and mm-wave regions to achieve the required data rates and bandwidth targets [4–7]. In this portion of the spectrum, the 28 GHz band has attracted a great deal of attention from the research community due to the low atmospheric attenuation, which is one of the most important and non-ignorable issues in mm-wave communication [8,9]. Because these attenuations make the signal weak when traveling from the transmitter to the receiver side, a low-quality and weak strength signal is received by the user [10,11]. A high gain antenna with a Multiple Input Multiple Output (MIMO) configuration can resolve these issues and improve the user communication experience by using the multi-path property [12–15]. In the literature, several antennas have been

99

reported for mm-wave applications. A study [16] reports a 5G antenna with a MIMO configuration possessing a 23–40 GHz wide response. The area of the proposed design is noted to be 80×80 mm^2. Although a huge bandwidth is achieved by the proposed MIMO design, the size of it is observed to be quite large. Similarly, a circular, polarized MIMO antenna with four ports is presented in [17]. An antenna element is placed within a center of a large number of parasitic elements to improve the radiation characteristics; however, the complexity of the proposed MIMO design is increased due to the use of parasitic elements. In Reference [18], the authors present a single antenna element with an overall size of 10×6 mm^2 with a 6.59 dB maximum gain. Although the gain achieved by the antenna element is sufficient, the implementations of electromagnetic band-gap structures (EBG) make the coupling issues severe and the fabrication task difficult. Likewise, array antenna operation at two frequency bands—i.e., 28 GHz and 38 GHz—is proposed in [19]. The high gain of the reported antenna makes this option attractive, but it lacks the MIMO property. An 8×8 MIMO configuration with a total volume of $31.2 \times 31.2 \times 1.57$ mm^3 is proposed in [20], giving a resonance at 25.2 GHz with a 8.7 dB maximum gain, while large numbers of back and side lobes are observed in the proposed antenna radiation response. In Reference [21], an antenna for 4G and 5G applications is presented, giving a resonance of 28 GHz with a total antenna size of 75×85 mm^2. The gain yielded at the mm-wave band is 5.13 dB, while large numbers of back and side lobes are found in the radiation response, which make the proposed antenna undesirable for mm-wave communication. Similarly, a mm-wave antenna design with a T-shaped structure covering the frequency band approximately from 25.1 GHz to 37.5 GHz is proposed in [22], while a four-element antenna with an area of 30×35 mm^2 and a peak gain of 8.3 dB and a MIMO antenna working in the 28 GHz frequency band are proposed in [23,24], respectively.

With the announcement of 5G sub-6 GHz and mm-wave potential bands, the mm-wave 28 GHz band has been seen as a potential candidate for future mm-wave smartphones. Several studies [25–28] have reported reserved space for mm-wave antennas, and some have included mm-wave designs. In Reference [26], a multi-band four-port antenna system is reported with an overall length and width of 158×78 mm^2 with a gain of 5.1 dB at 28 GHz frequency. The mm-wave antenna generated three different resonances of 28 GHz, 37 GHz and 39 GHz. The four antennas are placed on the central edges of the board, but their individual dimensions limit them to a maximum of four elements. In Reference [27], a novel T slot four-element antenna system consisting of two element-feed networks is presented, with each resonating at two distinct bands: one below sub-6 GHz and the other in the mm-wave band from 24 GHz to 29 GHz with a peak gain of 5 dB at 28 GHz. Although the study reported stable radiation patterns and performance parameters, the ground plane assembly and larger dimensions of radiating elements prevent the structure from being used with higher numbers of elements. An eight-port tapered slot antenna system with overall dimensions of $104 \times 104 \times 0.501$ mm^3 is reported in [28]. The antenna exhibited dual-band frequency response on a lower frequency band of 2.45 GHz and higher mm-wave mode of 28 GHz, providing an isolation level of 16 dB among two radiating elements and an efficiency of 80%. The reported multi-band antenna system with a tapered slot increases the overall complexity of the system. In these studies, it is observed that mm-wave resonating structures are complex and large in dimension. As 5G mainly focuses on high data rates with less latency as compared to 4G systems, the mm-wave currently has a huge bandwidth to be occupied and to be used since the sub-6 GHz spectrum is already congested. From the reported literature, single element structures have been observed, as well as those in linear array forms without MIMO assemblies. This work focuses on developing an MIMO antenna with fewer coupling phenomena for 28 GHz mm-wave applications. Therefore, the development of novel-shaped antenna with a circular ring-shaped radiating structure for the mm-wave applications with high isolation without the addition of a defected ground slot (DGS), tapered slot or electromagnetic band gap (EBG) structures is presented. The proposed shape is transformed into an MIMO configuration of four elements, and good impedance bandwidth and gain is achieved. The proposed structure can easily be embedded in smart phones as it provides and satisfies pattern diversity and all essential MIMO characteristics.

2. Antenna Design

The proposed antenna design was printed on a 0.787 mm thick Rogers 5880 substrate with a relative permittivity of 2.2 and loss tangent of 0.0009. The front view of the single element is shown in Figure 1. The proposed design comprised five ring structures fused in such a way that a modified novel spiral shape was formed. The dimensions of proposed element were as follows: Circle 1 = 4 mm, Circle 2 = 1.5 mm and feed length = 10 mm with a feed width of 0.76 mm. The width of the circular slots was kept at 0.4 mm, and the distance between larger and smaller rings was also kept 0.4 mm. The length and width of the substrate was an LS of 14 mm and WS of 14 mm, while the length of the ground plane was kept at 7 mm only. The ground plane was incorporated with a half circle with an outer radius of 3 mm and inner radius of 2.6 mm, respectively.

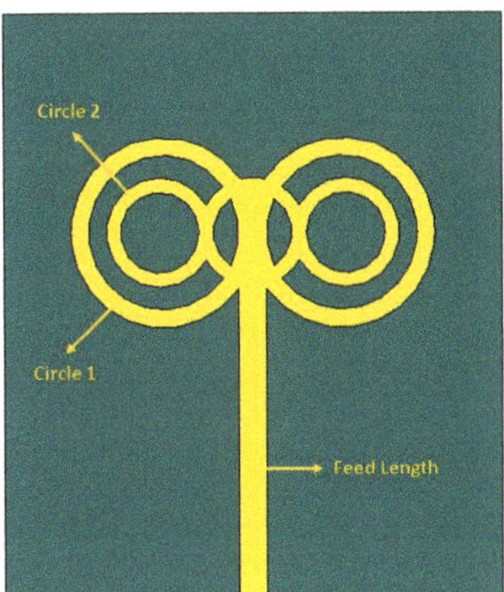

Figure 1. Proposed single-element antenna.

MIMO Configuration

The proposed structure was transformed into a four-port MIMO array with an overall length and width of 30 mm × 30 mm as shown in Figure 2. The MIMO assembly was arranged in such a way that the radiating elements were at a 90° distance from each other.

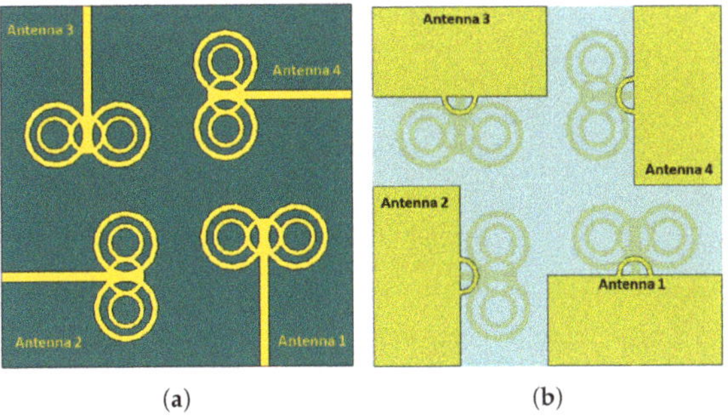

(**a**)   (**b**)

Figure 2. Multiple Input–Multiple Output (MIM) configuration. (**a**) View of the front; (**b**) view of the back.

Such a MIMO configuration has been found useful as it offers fewer coupling effects and provides much better spatial and pattern diversity. Copper with a very stable conductivity of 5.8×10^7 S/m was used for the radiating element. Due to the very stable conductivity of copper, its effect on the impedance matching was very low. The proposed design was modeled in Computer Simulation Technology CST (2017) under a higher meshing environment.

3. Results and Discussion

The proposed MIMO antenna was designed in CST microwave studio 2017 and fabricated using an LPKF (Leiterplatten–Kopierfrasen) machine. Figure 3 shows the proposed design fabricated model.

Figure 3. Fabricated prototype of MIMO array.

3.1. S Parameter Analysis

The S-parameter response is discussed in this section. The single-element reflection co-efficient response was tuned at 28 GHz with an impedance bandwidth of 2 GHz, as shown in Figure 4. In the MIMO configuration, the reflection co-efficient responses of Antenna 1 and Antenna 2 were slightly shifted but sufficiently covered the required bandwidth. Due to the perpendicular placement of radiating elements, the isolation of MIMO ports was found to be better than −29 dB. The simulated S-parameter response is shown in Figure 5a,b, while measured results are shown in Figure 5c,d. The measured coefficient results are slightly shifted, which can be attributed to human and measuring environment error. The measured isolation levels of the antenna were also found to be better than −29 dB levels, confirming the performance characteristics of the proposed structure.

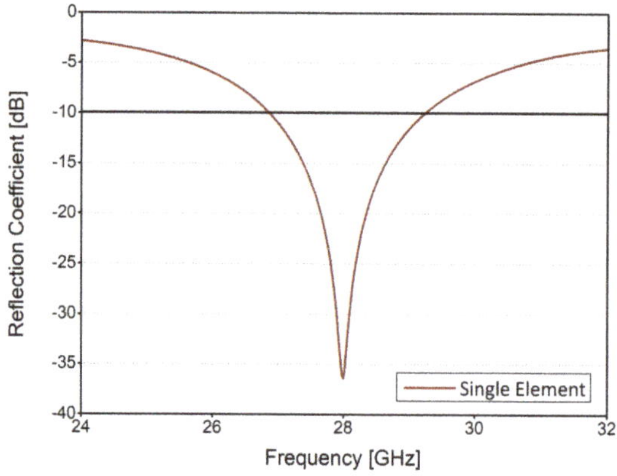

Figure 4. Simulated S-parameter of the single element.

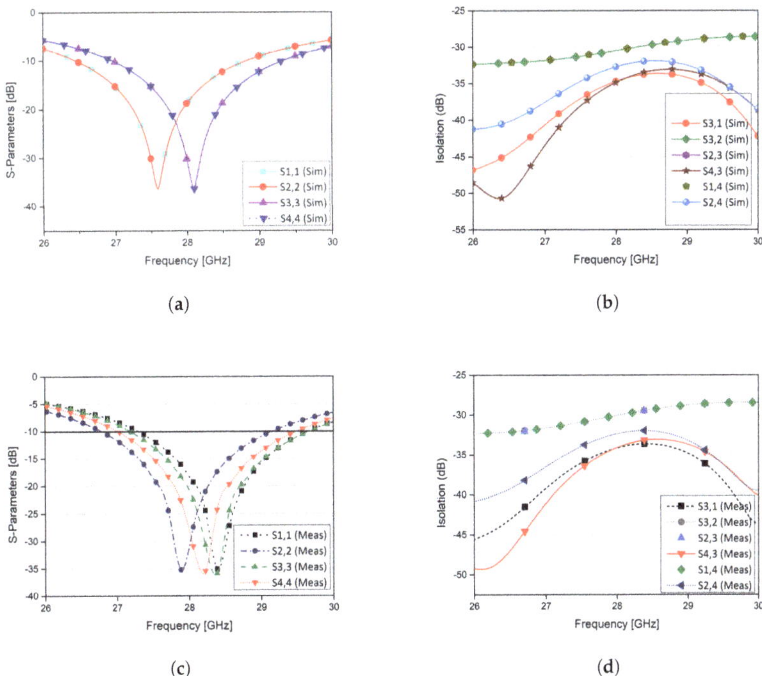

Figure 5. (**a**) Reflection coefficient of the proposed MIMO configuration, (**b**) simulated port isolation, (**c**) measured reflection coefficient of the proposed MIMO configuration, (**d**) measured port isolation.

3.2. Radiation Pattern

The simulated and measured radiation patterns were taken at a central frequency of 28 GHz on two principal planes, which were E and H planes, or at $\phi = 0°$ and $\phi = 90°$ as shown in Figure 6. The main lobe direction of antenna 1 in $\phi = 0°$ was between 90° and 150°, while at $\phi = 90°$, the main lobe was between 60° and 90°. Similarly, antenna 3 was the mirror image of antenna 1, and the radiation patterns of antenna 3 showed the main lobe between 210° and 270°; at $\phi = 90°$, the main lobe was between 240° and 300°. Regarding the radiation characteristics of antenna 2 and antenna 4, the main lobe was found to be in a reverse order to that of antenna 1 and antenna 3. This can be attributed to the perpendicular MIMO assembly of radiating elements, thus providing unique pattern diversity characteristics that are helpful for mitigating the multipath effect for communication systems. The variations in the simulated and measured radiation patterns can be attributed to the presence of attenuation and cable losses with probability of error in measuring the far-field setup. Although slight variation exists, the proposed MIMO structure gives a good impression of the simulated and measured radiation patterns. The antenna's efficiency is given in Figure 7, which clearly shows that proposed structure is highly efficient throughout the band of interest. Figure 8 shows the 3D radiation patterns of the proposed MIMO antenna system. The 3D figures shows that the proposed antenna system provides pattern diversity from each port.

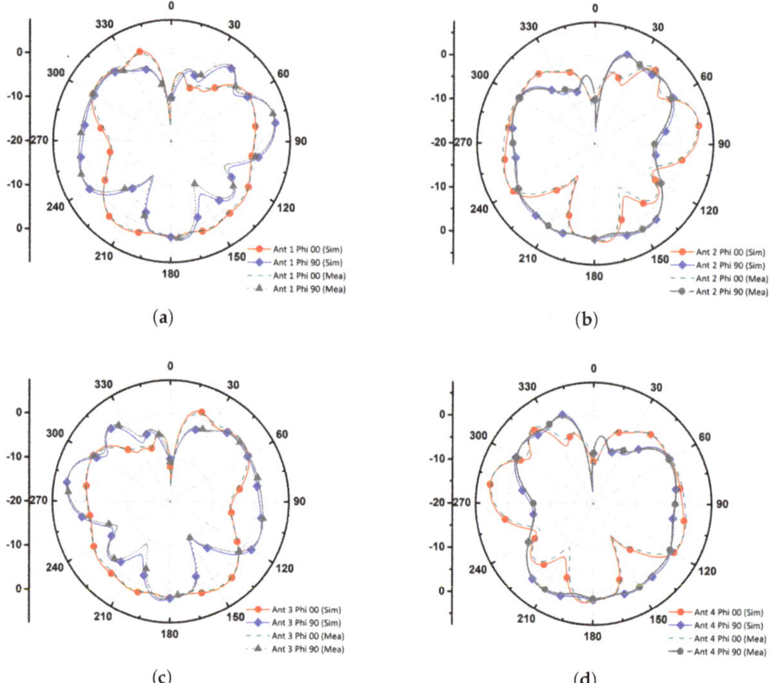

Figure 6. Simulated and measured radiation patterns: (**a**) AE-1 (**b**) AE-2 (**c**) AE-3 (**d**) AE-4. The polar coordinates show degrees, the y-axis is plotted in dBi.

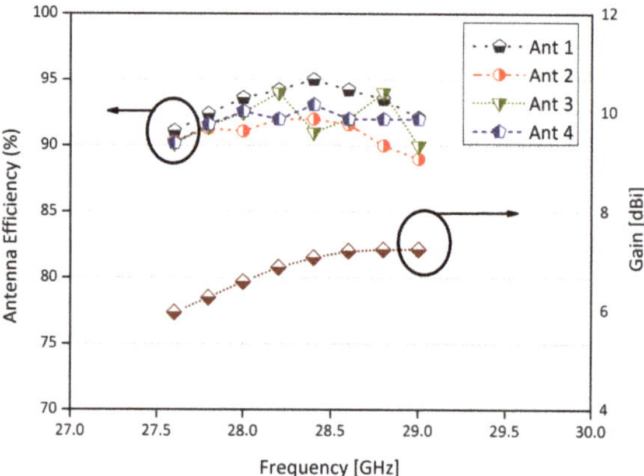

Figure 7. Antenna Efficiency and Simulated Gain of MIMO configuration.

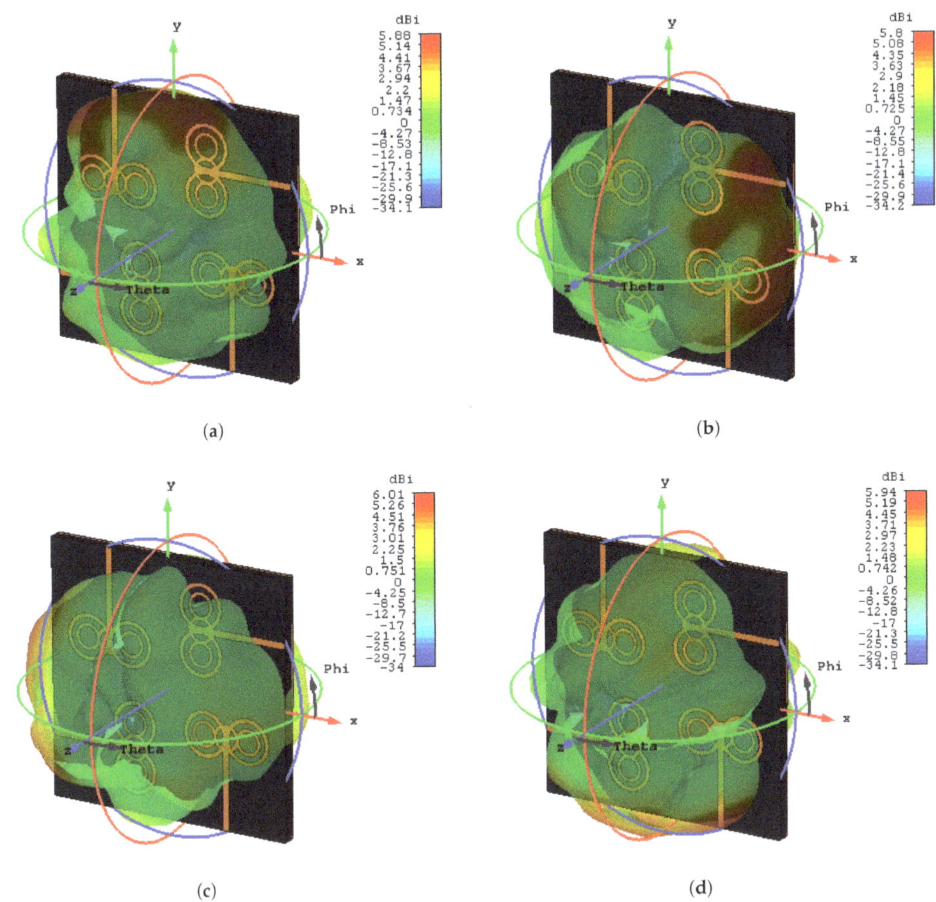

Figure 8. Simulated and measured radiation patterns (**a**) AE-1 (**b**) AE-2 (**c**) AE-3 (**d**) AE-4.

3.3. Surface Current Distributions

Figure 9 shows the distribution of the surface current for the proposed design operating at the 28 GHz band of interest for each antenna element, respectively. From the distribution of the surface current, it can be seen that the intensity of the current is high around the connected circle-shaped rings, showing that the contribution of these circular-shaped rings is more significant in the mm-wave band. Moreover, the level of the coupling among the elements is quite ignorable according to the distribution pattern of the current, which is achieved by the assembly of the elements with a shift of 90° per element.

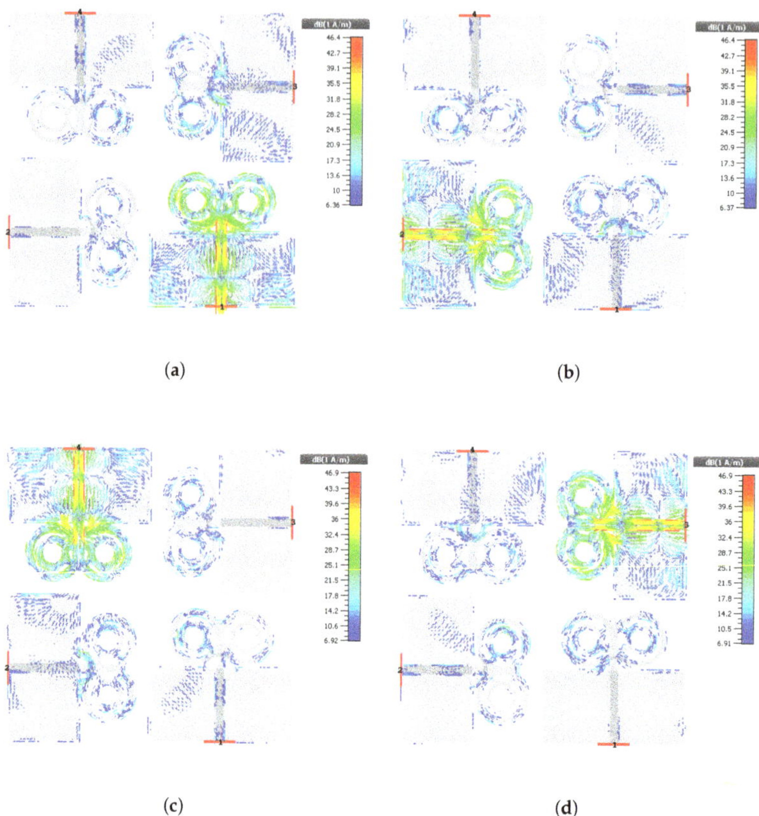

Figure 9. Distribution pattern of current at 28 GHz for (**a**) antenna 1, (**b**) antenna 2, (**c**) antenna 3, (**d**) antenna 4.

4. Mimo Parameters

4.1. Envelope Correlation Coefficient

To check the level of the coupling among elements of the proposed MIMO antenna, the envelope correlation coefficient (ECC) is evaluated. The S-parameter method is utilized to evaluate the response of the ECC [29]. A low value of the ECC—i.e., below 0.16—is observed in Figure 10 across the desired band, showing that the level of interference is low among the elements.

$$ECC = \frac{|\iint_{4\pi}(\vec{B}_i(\theta,\phi)) \times (\vec{B}_j(\theta,\phi))\,d\Omega|^2}{\iint_{4\pi}|(\vec{B}_i(\theta,\phi))|^2\,d\Omega \iint_{4\pi}|(\vec{B}_j(\theta,\phi))|^2\,d\Omega} \tag{1}$$

where $\vec{B}_i(\theta,\phi)$ denotes the 3D radiation pattern upon the excitation of the ith antenna and $\vec{B}_j(\theta,\phi)$ denotes the 3D radiation pattern upon the excitation of the jth antenna. Ω is the solid angle.

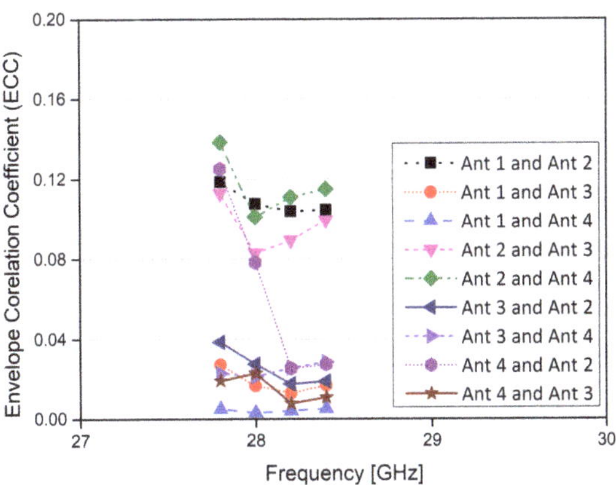

Figure 10. Envelope correlation coefficient for the MIMO antenna.

4.2. Mean Effective Gain (MEG)

The mean effective gain (MEG) is an important MIMO performance parameter which evaluates power imbalances in the MIMO system over a propagation environment. To fulfill the balance power standard and for optimal diversity performance with a good channel characteristic, the alteration of the MEGs of any two antennas in the MIMO system should be less than 3 dB. The calculated MEG value of the proposed MIMO antenna is shown in Table 1 and is calculated through the standard equation presented in [5].

$$MEG = \int_{-\pi}^{\pi} \int_{0}^{\pi} [\frac{r}{r+1} G_\theta(\theta,\phi) P_\theta(\theta,\phi) + \frac{1}{1+r} G_\phi(\theta,\phi) P_\phi(\theta,\phi)] sin\theta d\theta d\phi \qquad (2)$$

where $G_\phi(\theta,\phi)$ and $P_\theta(\theta,\phi)$ are angle of arrival and r is the cross polar ratio which can be expressed as Equation (3).

$$r = 10log_{10}(\frac{P_{vpa}}{P_{hpa}}) \qquad (3)$$

where the powers received by the vertically polarized antenna and horizontally polarized antenna are represented as P_{vpa} and P_{hpa}, respectively.

Table 1. Calculated mean effective gain (MEG) values of the MIMO antenna.

Frequency (GHz)	MEG1 (dB)	MEG2 (dB)	MEG3 (dB)	MEG4 (dB)
28	−4.1	−3.9	−3.2	−3.1
28.5	−3.09	−2.87	−3.5	−3.4
29	−2.9	−3.02	−2.8	−3.4

Table 2 provides published work comparison with proposed four port MIMO structure.

Table 2. Literature Comparison with the proposed MIMO antenna.

Ref.	Ports	Size of Antenna in mm (LxW)	Isolation (dB)	Eff. %	Gain (dBi)	ECC
[4]	6	136×68	-10	50	3.9	<0.3
[5]	8	150×75	-12.5	70	4.9	<0.15
[30]	4	55×110	-28.32	88.25	8.27	<0.047
[31]	3	130×80	-53	80	10.3	N/A
[32]	4	26.6×3.25	N/A	N/A	5.2	N/A
[29]	4	19.25×26	-25	86	5.72	<0.04
[33]	2	55×110	-27.3	N/A	9.49	<0.06
[34]	2	26×11	-25	98.6	5.7	<0.01
Proposed	4	30×30	<-29	92	6.1	<0.16

5. Conclusions

In this article, a novel infinity-shaped MIMO antenna is presented that resonates at 28 GHz in the mm-wave band for future 5G communication devices. The proposed MIMO antenna consists of four antenna elements which are arranged with a shift of 90°. Each antenna element radiating portion consists of four circle-shaped ring patches, which contribute mostly to achieve the operation in the desired frequency band. Furthermore, these four circle-shaped rings surround another circle located at the center; above the feed line in the case of each antenna element, respectively. The isolation achieved for the operating bandwidth is better than -29 dB, demonstrating low mutual coupling. Moreover, the peak gain and total efficiency obtained are 6.1 dBi and 92%, respectively, for the operating bandwidth of the proposed MIMO antenna. The lower value of ECC (<0.16), satisfactory efficiency and gain, good return loss and decent coherence between the simulated and measured results make the proposed design a competent candidate for future 5G-based mm-wave communication.

Author Contributions: Conceptualization, S.H.K. and M.M.K.; methodology, S.H.K., A.I. and A.A.; Software, S.H.K. and M.M.K.; Validation, A.I., S.I.S. and A.A.; Formal Analysis, A.A., S.I.S. and X.-c.R.; Investigation, S.H.K., M.R.A. and M.A.; Resources, A.I., M.A. and S.I.S.; Data Curation, S.H.K. and M.R.A.; Writing, original draft preparation, A.A. and X.R.; Writing, review and editing, S.Y. and M.M.K.; Project administration, X.-c.R., M.A. and S.Y.; Funding acquisition, X.-c.R. All authors have read and agreed to the published version of the manuscript.

Funding: This project was supported by the National Natural Science Foundation of China (61861043).

Data Availability Statement: No new data were created or analyzed in this study. Data sharing is not applicable to this article.

Conflicts of Interest: The authors declare no conflict of interest.

References

1. Varga, P.; Peto, J.; Franko, A.; Balla, D.; Haja, D.; Janky, F.; Soos, G.; Ficzere, D.; Maliosz, M.; Toka, L. 5g support for industrial iot applications–challenges, solutions, and research gaps. *Sensors* **2020**, *20*, 828. [CrossRef]
2. Hilt, A. Availability and Fade Margin Calculations for 5G Microwave and Millimeter-Wave Anyhaul Links. *Appl. Sci.* **2019**, *9*, 5240. [CrossRef]
3. Khalily, M.; Tafazolli, R.; Xiao, P.; Kishk, A.A. Broadband mm-wave microstrip array antenna with improved radiation characteristics for different 5G applications. *IEEE Trans. Antennas Propag.* **2018**, *66*, 4641–4647. [CrossRef]
4. Abdullah, M.; Kiani, S.H.; Abdulrazak, L.F.; Iqbal, A.; Bashir, M.; Khan, S.; Kim, S. High-performance multiple-input multiple-output antenna system for 5G mobile terminals. *Electronics* **2019**, *8*, 1090. [CrossRef]
5. Kiani, S.H.; Altaf, A.; Abdullah, M.; Muhammad, F.; Shoaib, N.; Anjum, M.R.; Damaševičius, R.; Blažauskas, T. Eight Element Side Edged Framed MIMO Antenna Array for Future 5G Smart Phones. *Micromachines* **2020**, *11*, 956. [CrossRef] [PubMed]
6. Pi, Z.; Khan, F. An introduction to millimeter-wave mobile broadband systems. *IEEE Commun. Mag.* **2011**, *49*, 101–107. [CrossRef]
7. Ullah, H.; Tahir, F.A. A Novel Snowflake Fractal Antenna for Dual-Beam Applications in 28 GHz Band. *IEEE Access* **2020**, *8*, 19873–19879. [CrossRef]

8.   NetWorld2020 ETP. 5G: Challenges, Research Priorities, and Recommendations; Joint White Paper. September 2014. Available online: https://networld2020.eu/wp-content/uploads/2015/01/Joint-Whitepaper-V12-clean-after-consultation.pdf (accessed on 6 January 2021).

9.   Shayea, I.; Rahman, T.A.; Azmi, M.H.; Islam, M.R. Real measurement study for rain rate and rain attenuation conducted over 26 GHz microwave 5G link system in Malaysia. *IEEE Access* **2018**, *6*, 19044–19064. [CrossRef]

10.   Aliakbari, H.; Abdipour, A.; Costanzo, A.; Masotti, D.; Mirzavand, R.; Mousavi, P. Performance investigation of space diversity for a 28/38 GHz MIMO antenna (applicable to mm-wave mobile network). In Proceedings of the 2016 Fourth International Conference on Millimeter-Wave and Terahertz Technologies (MMWaTT), Tehran, Iran, 20–22 December 2016; pp. 41–44.

11.   Sulyman, A.I.; Alwarafy, A.; MacCartney, G.R.; Rappaport, T.S.; Alsanie, A. Directional radio propagation path loss models for millimeter-wave wireless networks in the 28-, 60-, and 73-GHz bands. *IEEE Trans. Wirel. Commun.* **2016**, *15*, 6939–6947. [CrossRef]

12.   Zhang, J.; Ge, X.; Li, Q.; Guizani, M.; Zhang, Y. 5G millimeter-wave antenna array: Design and challenges. *IEEE Wirel. Commun.* **2016**, *24*, 106–112. [CrossRef]

13.   Wang, F.; Duan, Z.; Wang, X.; Zhou, Q.; Gong, Y. High Isolation Millimeter-Wave Wideband MIMO Antenna for 5G Communication. *Int. J. Antennas Propag.* **2019**, *2019*, 4283010. [CrossRef]

14.   Guo, J.; Cui, L.; Li, C.; Sun, B. Side-edge frame printed eight-port dual-band antenna array for 5G smartphone applications. *IEEE Trans. Antennas Propag.* **2018**, *66*, 7412–7417. [CrossRef]

15.   Yang, B.; Yu, Z.; Dong, Y.; Zhou, J.; Hong, W. Compact tapered slot antenna array for 5G millimeter-wave massive MIMO systems. *IEEE Trans. Antennas Propag.* **2017**, *65*, 6721–6727. [CrossRef]

16.   Sehrai, D.A.; Abdullah, M.; Altaf, A.; Kiani, S.H.; Muhammad, F.; Tufail, M.; Irfan, M.; Glowacz, A.; Rahman, S. A Novel High Gain Wideband MIMO Antenna for 5G Millimeter Wave Applications. *Electronics* **2020**, *9*, 1031. [CrossRef]

17.   Hussain, N.; Jeong, M.J.; Abbas, A.; Kim, N. Metasurface-based single-layer wideband circularly polarized MIMO antenna for 5G millimeter-wave systems. *IEEE Access* **2020**, *8*, 130293–130304. [CrossRef]

18.   Khan, J.; Sehrai, D.A.; Khan, M.A.; Khan, H.A.; Ahmad, S.; Ali, A.; Arif, A.; Memon, A.A.; Khan, S. Design and Performance Comparison of Rotated Y-Shaped Antenna Using Different Metamaterial Surfaces for 5G Mobile Devices. *CMC-Comput. Mater. Contin.* **2019**, *60*, 409–420. [CrossRef]

19.   Khan, J.; Sehrai, D.A.; Ali, U. Design of dual band 5G antenna array with SAR analysis for future mobile handsets. *J. Electr. Eng. Technol.* **2019**, *14*, 809–816. [CrossRef]

20.   Shoaib, N.; Shoaib, S.; Khattak, R.Y.; Shoaib, I.; Chen, X.; Perwaiz, A. MIMO antennas for smart 5G devices. *IEEE Access* **2018**, *6*, 77014–77021. [CrossRef]

21.   Iffat Naqvi, S.; Hussain, N.; Iqbal, A.; Rahman, M.; Forsat, M.; Mirjavadi, S.S.; Amin, Y. Integrated LTE and Millimeter-Wave 5G MIMO Antenna System for 4G/5G Wireless Terminals. *Sensors* **2020**, *20*, 3926. [CrossRef]

22.   Jilani, S.F.; Alomainy, A. Millimetre-wave T-shaped antenna with defected ground structures for 5G wireless networks. In Proceedings of the 2016 Loughborough Antennas & Propagation Conference (LAPC), Loughborough, UK, 14–16 November 2016; pp. 1–3.

23.   Khalid, M.; Iffat Naqvi, S.; Hussain, N.; Rahman, M.; Mirjavadi, S.S.; Khan, M.J.; Amin, Y. 4-Port MIMO antenna with defected ground structure for 5G millimeter wave applications. *Electronics* **2020**, *9*, 71. [CrossRef]

24.   Wani, Z.; Abegaonkar, M.P.; Koul, S.K. A 28-GHz antenna for 5G MIMO applications. *Prog. Electromagn. Res.* **2018**, *78*, 73–79. [CrossRef]

25.   Ojaroudi Parchin, N.; Jahanbakhsh Basherlou, H.; Alibakhshikenari, M.; Ojaroudi Parchin, Y.; Al-Yasir, Y.I.; Abd-Alhameed, R.A.; Limiti, E. Mobile-phone antenna array with diamond-ring slot elements for 5G massive MIMO systems. *Electronics* **2019**, *8*, 521. [CrossRef]

26.   Al Abbas, E.; Ikram, M.; Mobashsher, A.T.; Abbosh, A. MIMO Antenna System for Multi-Band Millimeter-Wave 5G and Wideband 4G Mobile Communications. *IEEE Access* **2019**, *7*, 181916–181923. [CrossRef]

27.   Naqvi, S.I.; Naqvi, A.H.; Arshad, F.; Riaz, M.A.; Azam, M.A.; Khan, M.S.; Amin, Y.; Loo, J.; Tenhunen, H. An integrated antenna system for 4G and millimeter-wave 5G future handheld devices. *IEEE Access* **2019**, *7*, 116555–116566. [CrossRef]

28.   Ikram, M.; Nguyen-Trong, N.; Abbosh, A. Multiband MIMO microwave and millimeter antenna system employing dual-function tapered slot structure. *IEEE Trans. Antennas Propag.* **2019**, *67*, 5705–5710. [CrossRef]

29.   Tu, D.T.T.; Thang, N.G.; Ngoc, N.T.; Phuong, N.T.B.; Van Yem, V. 28/38 GHz dual-band MIMO antenna with low mutual coupling using novel round patch EBG cell for 5G applications. In Proceedings of the 2017 International Conference on Advanced Technologies for Communications (ATC), Quy Nhon, Vietnam, 18–20 October 2017; pp. 64–69.

30.   Marzouk, H.M.; Ahmed, M.I.; Shaalan, A.H.A.M. Novel dual-band 28/38 GHz MIMO antennas for 5G mobile applications. *Prog. Electromagn. Res.* **2019**, *93*, 103–117. [CrossRef]

31.   Liu, Y.; Li, Y.; Ge, L.; Wang, J.; Ai, B. A Compact Hepta-Band Mode-Composite Antenna for Sub (6, 28, and 38) GHz Applications. *IEEE Trans. Antennas Propag.* **2020**, *68*, 2593–2602. [CrossRef]

32.   Sunthari, P.M.; Veeramani, R. Multiband microstrip patch antenna for 5G wireless applications using MIMO techniques. In Proceedings of the 2017 First International Conference on Recent Advances in Aerospace Engineering (ICRAAE), Coimbatore, India, 3–4 March 2017; pp. 1–5.

33. Marzouk, H.; Ahmed, M.; Shaalan, A. A Novel Dual-band 28/38 GHz Slotted Microstip MIMO Antenna for 5G Mobile Applications. In Proceedings of the 2019 IEEE International Symposium on Antennas and Propagation and USNC-URSI Radio Science Meeting, Atlanta, GA, USA, 7–12 July 2019; pp. 607–608.
34. Ali, W.; Das, S.; Medkour, H.; Lakrit, S. Planar dual-band 27/39 GHz millimeter-wave MIMO antenna for 5G applications. *Microsyst. Technol.* **2020**, 1–10. [CrossRef]

Article

Theoretical Study of the Input Impedance and Electromagnetic Field Distribution of a Dipole Antenna Printed on an Electrical/Magnetic Uniaxial Anisotropic Substrate

Mohamed Lamine Bouknia [1], Chemseddine Zebiri [1], Djamel Sayad [2], Issa Elfergani [3,4,*], Jonathan Rodriguez [3,5], Mohammad Alibakhshikenari [6,*], Raed A. Abd-Alhameed [4], Francisco Falcone [7] and Ernesto Limiti [6]

[1] Laboratoire d'Electronique de Puissance et Commande Industrielle (LEPCI), Department of Electronics, University of Ferhat Abbas, Sétif -1-, Sétif 19000, Algeria; ml.bouknia@univ-setif.dz (M.L.B.); czebiri@univ-setif.dz (C.Z.)

[2] Laboratoire d'Electrotechnique de Skikda (LES), University 20 Aout 1955-Skikda, Skikda 21000, Algeria; d.sayad@univ-skikda.dz

[3] Instituto de Telecomunicações, Campus Universitário de Santiago, 3810-193 Aveiro, Portugal; jonthan@av.it.pt

[4] School of Electrical Engineering and Computer Science, University of Bradford, Bradford BD71DP, UK; r.a.a.abd@bradford.ac.uk

[5] Faculty of Computing, Engineering and Science, University of South Wales, Pontypridd CF37 1DL, UK

[6] Electronic Engineering Department, University of Rome "Tor Vergata", Via Del Politecnico 1, 00133 Rome, Italy; Limiti@ing.uniroma2.it

[7] Electric, Electronic and Communication Engineering Department, Public University of Navarre, 31006 Pamplona, Spain; francisco.falcone@unavarra.es

* Correspondence: i.t.e.elfergani@av.it.pt or i.elfergani@bradford.ac.uk (I.E.); alibakhshikenari@ing.uniroma2.it (M.A.); Tel.: +351-234-377900 (I.E.)

Citation: Bouknia, M.L.; Zebiri, C.; Sayad, D.; Elfergani, I.; Rodriguez, J.; Alibakhshikenari, M.; Abd-Alhameed, R.A.; Falcone, F.; Limiti, E. Theoretical Study of the Input Impedance and Electromagnetic Field Distribution of a Dipole Antenna Printed on an Electrical/Magnetic Uniaxial Anisotropic Substrate. *Electronics* **2021**, *10*, 1050. https://doi.org/10.3390/electronics10091050

Academic Editor: Dong Ho Cho

Received: 31 March 2021
Accepted: 26 April 2021
Published: 29 April 2021

Publisher's Note: MDPI stays neutral with regard to jurisdictional claims in published maps and institutional affiliations.

Abstract: The present work considers the investigation of the effects of both electrical and magnetic uniaxial anisotropies on the input impedance, resonant length, and fields distribution of a dipole printed on an anisotropic grounded substrate. In this study, the associated integral equation, based on the derivation of the Green's functions in the spectral domain, is numerically solved employing the method of moments. In order to validate the computing method and the evaluated calculation code, numerical results are compared with available data in the literature treating particular cases of electrical uniaxial anisotropy; reasonable agreements are reported. Novel results of the magnetic uniaxial anisotropy effects on the input impedance and the evaluated electromagnetic field are presented and discussed. This work will serve as a stepping stone for further works for a better understanding of the electromagnetic field behavior in complex anisotropic and bi-anisotropic media.

Keywords: green's functions; method of moments; uniaxial anisotropy; input impedance; field distributions; dipole antenna

1. Introduction

Since a few decades, theoretical and experimental studies of the interaction of electromagnetic waves with complex-media structures, anisotropic or bianisotropic, have been extensively investigated for their innovative applications, including geophysical explorations, communications with buried and submerged antennas, microwave/millimeter integrated circuits, optical devices, etc. [1–7].

In general, complex media have received increasing interest from both scientists and researchers in the context of artificial media with new and interesting properties due to their additional degree of freedom [8]. Anisotropy is an intrinsic property found in crystals, layered structures, composite materials, and other natural materials, in addition to artificial materials. The effect of anisotropy must be considered and cannot be ignored in the prediction of unusual properties in engineering designs such as for sensing and antenna applications [9–11]. They have attracted much interest and support from researchers and

111

manufacturers as powerful instruments with interesting growth potential in microwave applications [12]. Many studies have been performed to characterize microwave structures printed on complex media, ferrites, metamaterials, chiral by employing numerical and analytical methods [11–24]. In [11], a detailed analytical model is derived for the circularly polarized slot antenna, built on a ferrite substrate. This model is based on an integral equation for the radial electric field on the slot. In [12], a patch antenna model is first simulated using the finite element method-based HFSS software and then fabricated on a multiple-ferrite-cored substrate. Recently, in [13], Karma et al. studied microstrip transmission lines with anisotropic and uniaxial anisotropic substrates using the discrete mode matching method. A technique for the calculation of the input impedance of a microstrip antenna printed on chiral substrate based on the integral equation with the Cauchy singularity is derived in [14]. This technique is used in [15] to investigate the dependences of the input impedance, the magnitude and phase of the electric-field components on the radiator length for different types of chiral substrates. In [16], a method based on the volume integral equation (VIE) is used to evaluate the electromagnetic (EM) fields scattered by general anisotropic multilayer structures. In [17,18], for a microstrip patch antenna, the effect of a bianisotropic gyro-chiral substrate on the complex resonant frequency, half-power bandwidth and input impedance, and on the surface waves is presented, respectively. The analysis is based on the full-wave spectral method of moments using sinusoidal basis functions. In [19], the far field radiation of a Hertzian dipole for two-layered uniaxial anisotropic medium is investigated using the spectral method of moments based on the derivation of the spectral dyadic Green's functions (DGFs) to examine the effect of anisotropy, effect of layer thickness, and effect of dipole location on the radiation fields. The effect of the dielectric constitutive parameters on the input impedance and resonance lengths of a dipole antenna based on stratified electrical anisotropic and chiral substrates and a microstrip patch have been analyzed in [20–23] using the spectral method of moments [24]. Many attempts have been made to analyze the problem of dipole antennas on an anisotropic uniaxial media with the optical axis perpendicular to the substrate plane [25–29] and with arbitrary optical axes orientations [30]. The input impedance and mutual coupling of single and multilayer dipole antennas printed on isotropic, anisotropic, and chiral materials have been studied in [29,31–34].

This work extends past research works on printed dipoles by studying a planar dipole of arbitrary length printed on a uniaxial anisotropic dielectric structure. The electromagnetic field distributions and input impedance are examined by applying the spectral method of moments based on the immittance functions derivation [35–38]. The Method of Moments in the spectral domain is found to be a powerful numerical technique to solve integral equations [39], and considered as rigorous and full-wave numerical technique for solving open boundary electromagnetic problems and the employing of this method becomes an increasingly important research issue [4–10,17–24,28–31,34–38]. To accurately predict the electromagnetic behavior of microwave components, the method of the moment is widely used in the spectral domain [8–10,17,18,20]. The efficient spectral Galerkin-based method of moments (SGMoM) is extensively used to analyze microwave planar structures [34], such as microstrip lines and antennas, with perfect conductors, it was first applied in the microwave field by Harrington in 1968 [39,40]. To accelerate convergence and improve the computation efficient of this method, several procedures have been introduced [10,34,41]. Resolving for the resonant frequency and input impedance of the printed dipole will directly show how each individual element of the anisotropic tensor affects the characteristics of the printed dipole, in particular, the input impedance and the electric and magnetic field distributions. This will provide new and useful information on how to integrate the individual anisotropic permittivity and permeability elements into the design of printed dipoles. Clifford M. Krowne determined the magnetic and electric field distributions, as well as the Poynting vectors, for cross-section (RHM/LHM) of a microstrip guided wave structures [4,5], and for ferrite microstrip guided-wave structures [6,7].

Sections 2 and 3 covers the basic equations, formulations and solution of differential equations, Green's functions, and components of electric and magnetic fields expressions. The study begins with the examination of the effects of introducing anisotropy via a permittivity and permeability tensor. The electromagnetic field distributions have been obtained and compared with the isotropic case for different anisotropic cases.

2. Analytical Formulation

Uniaxial materials have the same element in two dimensions, permittivity or permeability, and different in the third dimension. The axis along the single value direction is called the optical axis. Figure 1 shows the structure considered in this analysis. The presented configuration will be considered to establish the electric and magnetic field distributions and determine how the input impedance is affected by the uniaxial anisotropic substrate. In this analysis, the uniaxial permittivity and permeability anisotropy is given by the following expressions, respectively:

$$[\varepsilon] = \varepsilon_0 \begin{bmatrix} \varepsilon_t & 0 & 0 \\ 0 & \varepsilon_t & 0 \\ 0 & 0 & \varepsilon_z \end{bmatrix} \tag{1a}$$

$$[\mu] = \mu_0 \begin{bmatrix} \mu_t & 0 & 0 \\ 0 & \mu_t & 0 \\ 0 & 0 & \mu_z \end{bmatrix} \tag{1b}$$

$[\varepsilon]$ and $[\mu]$ are, respectively, the tensors of the relative permittivity and permeability, which are assumed to be low-frequency dispersive in the microwave frequency band.

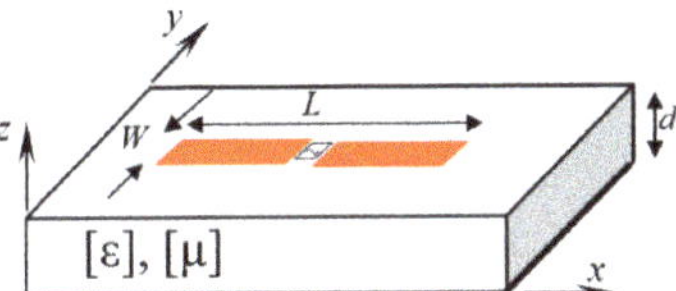

Figure 1. Printed dipole on uniaxial electrical- and magnetic- anisotropic substrate.

First, assuming the temporal dependence $e^{j\omega t}$ of the fields and applying Maxwell's equations in the Fourier domain with $\partial/\partial x \equiv -j\alpha$ and $\partial/\partial y \equiv -j\beta$ assumptions, we derive the expressions of the transverse components of the electromagnetic field $\tilde{E}_z$ and $\tilde{H}_z$, two decoupled homogeneous differential wave equations of the second order are obtained:

$$\frac{\partial^2 \tilde{E}_z}{\partial z^2} - \gamma_e^2 \tilde{E}_z = 0 \tag{2a}$$

$$\frac{\partial^2 \tilde{H}_z}{\partial z^2} - \gamma_h^2 \tilde{H}_z = 0 \tag{2b}$$

The dispersion relations are found to be as follows:

$$\gamma_e = \sqrt{\left(\frac{\varepsilon_t}{\varepsilon_z}(\alpha^2 + \beta^2) - \kappa_0^2 \varepsilon_t \mu_t \right)} \tag{2c}$$

$$\gamma_h = \sqrt{\left(\frac{\mu_t}{\mu_z}(\alpha^2 + \beta^2) - \kappa_0^2 \varepsilon_t \mu_t \right)} \tag{2d}$$

γ_e^2 and γ_h^2 represent the propagation constants of the TM and TE transverse modes, respectively. κ_0 is the free space wavenumber and ω is the angular frequency. α and β are the Fourier variables corresponding to the space domain wavenumbers κ_x and κ_y.

3. Method of Solution

Extensive mathematical manipulations of Maxwell's equations result in the wave Equations (2a) and (2b), which admit a general solution of the form (3a) and (3b). On the other hand, for the region above the substrate (air region), the spectral components are decreasing waves with z, for which the solutions (4a) and (4b) are assumed.

$$\tilde{E}_z(\gamma_e, z) = A_e \cosh(\gamma_e z) + B_e \sinh(\gamma_e z) \tag{3a}$$

$$\tilde{H}_z(\gamma_h, z) = A_h \sinh(\gamma_h z) + B_h \cosh(\gamma_h z) \tag{3b}$$

with A_e, B_e, A_h and B_h are complex constants

$$\tilde{E}_z(\gamma_0, z) = C_e e^{-\gamma_0(z-d)} \tag{4a}$$

$$\tilde{H}_z(\gamma_0, z) = C_h e^{-\gamma_0(z-d)} \tag{4b}$$

and

$$\gamma_0 = \sqrt{(\alpha^2 + \beta^2) - \kappa_0^2} \tag{4c}$$

and C_e and C_h are complex constants.

The boundary conditions have been applied to obtain the dyadic Green's functions. In the configuration of narrow dipoles, it is assumed that the width of the strip is very small, thus, the transverse current density in the y-direction is usually neglected [31]. Therefore, only the expression of $\tilde{G}_{xx}$ is presented:

$$\tilde{G}_{xx} = \frac{-j}{\omega \varepsilon_0 (\alpha^2 + \beta^2)} \left[\frac{\alpha^2 \gamma_0 \gamma_e}{(\gamma_0 \varepsilon_t \coth(\gamma_e d) + \gamma_e)} - \frac{\beta^2 \kappa_0^2 \mu_t}{(\gamma_h \coth(\gamma_h d) + \mu_t \gamma_0)} \right] \tag{5}$$

Mathematical development of the resulting equations lead to the formulation of the estimated electric field at the interface between the two regions as functions of the current densities $\tilde{J}_x$ and $\tilde{J}_y$. The expressions of the x-, y- and z-components of the electric and magnetic fields in the 1st and 2nd regions can be expressed as follows:

1st region:

$$\tilde{E}_{x1}(\alpha, \beta, z) = \frac{j}{\alpha^2 + \beta^2} \frac{1}{\omega \varepsilon_0} \left(-\alpha \gamma_e^2 \gamma_0 \mathrm{Se} \times A_e + \beta \kappa_0^2 \mu_t \mathrm{Sh} \times A_h \right) \tag{6a}$$

$$\tilde{E}_{y1}(\alpha, \beta, z) = \frac{j}{\alpha^2 + \beta^2} \frac{1}{\omega \varepsilon_0} \left(-\beta \gamma_e^2 \gamma_0 \mathrm{Se} \times A_e - \alpha \kappa_0^2 \mu_t \mathrm{Sh} \times A_h \right) \tag{6b}$$

$$\tilde{E}_{z1}(\alpha, \beta, z) = -\frac{\gamma_0 \gamma_{ec} \varepsilon_t}{\omega \varepsilon_0 \varepsilon_z} \mathrm{Se} \times A_e \tag{6c}$$

$$\tilde{H}_{x1}(\alpha, \beta, z) = \frac{1}{\alpha^2 + \beta^2} \left(\beta \gamma_0 \varepsilon_t \gamma_{ec} \mathrm{Se} \times A_e - \alpha \gamma_{hc} \mathrm{Sh} \times A_h \right) \tag{6d}$$

$$\tilde{H}_{y1}(\alpha, \beta, z) = \frac{1}{\alpha^2 + \beta^2} \left(-\alpha \gamma_0 \varepsilon_t \gamma_{ec} \mathrm{Se} \times A_e - \beta \gamma_{hc} \mathrm{Sh} \times A_h \right) \tag{6e}$$

$$\tilde{H}_{z1}(\alpha, \beta, z) = j \frac{\mu_t}{\mu_z} \mathrm{Sh} \times A_h \tag{6f}$$

2nd region:

$$\tilde{E}_{x2}(\alpha, \beta, z) = j \frac{e^{-\gamma_0(z-d)}}{\alpha^2 + \beta^2} \frac{1}{\omega \varepsilon_0} \left[-\alpha \gamma_0 \gamma_e^2 A_e + \beta \mu_t \kappa_0^2 A_h \right] \tag{7a}$$

$$\tilde{E}_{y2}(\alpha, \beta, z) = j \frac{e^{-\gamma_0(z-d)}}{\alpha^2 + \beta^2} \frac{1}{\omega \varepsilon_0} \left[-\beta \gamma_0 \gamma_e^2 A_e - \alpha \mu_t \kappa_0^2 A_h \right] \tag{7b}$$

$$\tilde{E}_{z2}(\alpha, \beta, z) = \frac{\gamma_e^2}{\omega \varepsilon_0} A_e e^{-\gamma_0(z-d)} \tag{7c}$$

$$\tilde{H}_{x2}(\alpha, \beta, z) = \frac{e^{-\gamma_0(z-d)}}{\alpha^2 + \beta^2}\left(-\beta\gamma_e^2 A_e + \mu_t \alpha \gamma_0 A_h\right) \tag{7d}$$

$$\tilde{H}_{y2}(\alpha, \beta, z) = \frac{e^{-\gamma_0(z-d)}}{\alpha^2 + \beta^2}\left(\alpha\gamma_e^2 A_e + \beta\mu_t \gamma_0 A_h\right) \tag{7e}$$

$$\tilde{H}_{z2}(\alpha, \beta, z) = j\mu_t A_h e^{-\gamma_0(z-d)} \tag{7f}$$

where $\tilde{J}_x$ and $\tilde{J}_y$ are the Fourier transforms of the current densities, and

$$A_e = \frac{\alpha\tilde{J}_x + \beta\tilde{J}_y}{\left(\gamma_e^2 + \gamma_0\varepsilon_t\gamma_e\mathrm{coth}(\gamma_e d)\right)} \tag{8a}$$

$$A_h = \frac{\beta\tilde{J}_x - \alpha\tilde{J}_y}{\left(\gamma_h\mathrm{coth}(\gamma_h d) + \gamma_0\mu_t\right)} \tag{8b}$$

$$\gamma_{ec} = \gamma_e\mathrm{coth}(\gamma_e d) \tag{8c}$$

$$\gamma_{hc} = \gamma_h\mathrm{coth}(\gamma_h d) \tag{8d}$$

$$Se = \frac{\mathrm{sinh}(\gamma_e z)}{\mathrm{sinh}(\gamma_e d)} \tag{8e}$$

$$Sh = \frac{\mathrm{sinh}(\gamma_h z)}{\mathrm{sinh}(\gamma_h d)} \tag{8f}$$

4. Fields Computations

The spectral domain immittance functions are used to evaluate the configuration shown in Figure 1. In particular, we seek to determine the distribution of the electric and magnetic fields on a printed dipole embedded on an anisotropic layer. The dipole has length L and width W. The electromagnetic field components are deduced after satisfying the boundary conditions on the printed dipole.

As mentioned in the section above, $\tilde{G}_{xx}$ represents the Green's function in the spectral domain and the Green's function in the spatial domain is merely the inverse Fourier transform. The components of the electric and magnetic fields in each region can be written in the spatial domain as [42,43]:

$$\Phi(x, y, z) = \frac{1}{(2\pi)^2}\int_{-\infty}^{\infty}\int_{-\infty}^{\infty}\tilde{\Phi}(\alpha, \beta, z)e^{j(\alpha x + \beta y)}d\alpha d\beta \tag{9}$$

With Formulation (9), a specific structure was chosen to be examined. The electromagnetic field expressions are derived in the spatial domain via the inverse Fourier transform. The Matlab® software [44] is used to plot the fields distributions.

5. Numerical Results

In this work, we are firstly interested in the input impedance, the resonant length of the dipole and secondly in the distribution of the electromagnetic fields. Before discussing the results obtained for the uniaxial anisotropy case, a validation of the calculation code, elaborated in Matlab, is carried out through a comparison with published literature.

5.1. Validation

In this subsection, we consider the basic configuration of a monolayer $0.1060\lambda_0$ thick substrate planar dipole antenna. The objective of this work is to analyze the effects of different electromagnetic parameters of the anisotropic substrate on the input impedance of the dipole, in addition to the electromagnetic field evaluation through the plotting of the electric and magnetic field distributions in the three principal planes XY, XZ, and YZ.

Calculations of a dipole structure on a uniaxial anisotropic structure have been performed. The results have been successfully compared with published ones. We initially considered the isotropic and uniaxial anisotropic cases ($\varepsilon_t = \varepsilon_z = 3.25$ and $\mu_t = \mu_z = 1$) and ($\varepsilon_t = 3.14$, $\varepsilon_z = 5.12$ and $\mu_t = \mu_z = 1$), respectively. Figure 2 depicts the input impedance (real and imaginary parts) of a planar dipole of width $W = 0.0004\lambda_0$ as a function of normalized length L/λ_0. These results represent a validation step of the accuracy of our calculations for both isotropic and anisotropic substrates. The representation shows good agreement with the data reported in [29]. In [29], only cases of electrical anisotropy were considered and no discussion of the effect of this component was conducted.

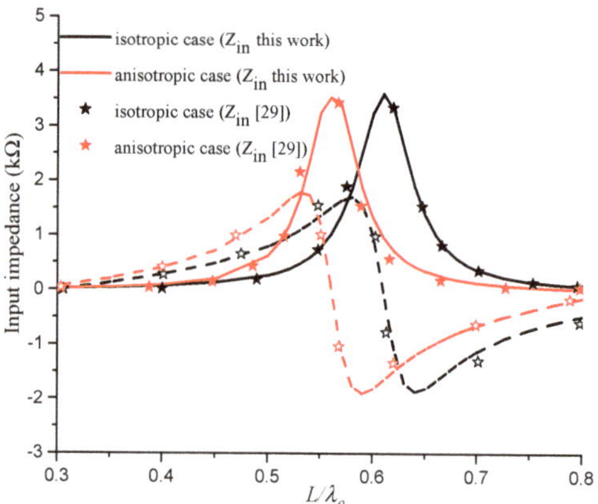

Figure 2. Input impedance of the dipole printed on isotropic and anisotropic layers.

5.2. Electromagnetic-Field Distributions in Isotropic Case

Once the expressions for the transverse and longitudinal components are found, the electromagnetic fields expressions are evaluated. Since the processing takes place in the spectral domain, the final fields must be brought back into the spatial domain. The Matlab functions *"quiver3"* and *"contour"* are used in plotting the field lines in their arrow and equiphase contour forms. It is clear from these results that the considered antenna structure shows the well-known shape of a dipole radiated electromagnetic field. The anisotropic results of the tangential-field lines are plotted for different planes at the equivalent resonant dipole length compared to the isotropic case. The magnitude of the electromagnetic fields are normalized with respect to the isotropic case for comparison.

In our case, we evaluate the electric and magnetic field distributions in a dipole antenna structure based on media with various uniaxial anisotropies and the effect of the different elements of the constitutive parameters. This will serve as a platform for the treatment of other cases of microwave structures based on more complex media such as: electric negative material (ENG), magnetic negative material (MNG), double negative material (DNG), and chirality.

Figure 3 shows the electric and magnetic fields arrow plots for the tangential fields components E_t and H_t, in the transverse plane with respect to z-, y- and x-axis, respectively, for the isotropic case. The arrow indicates the cross-sectional field vector direction and the arrow length designates the field magnitude, and the lines indicates the equi-phase field contour forms.

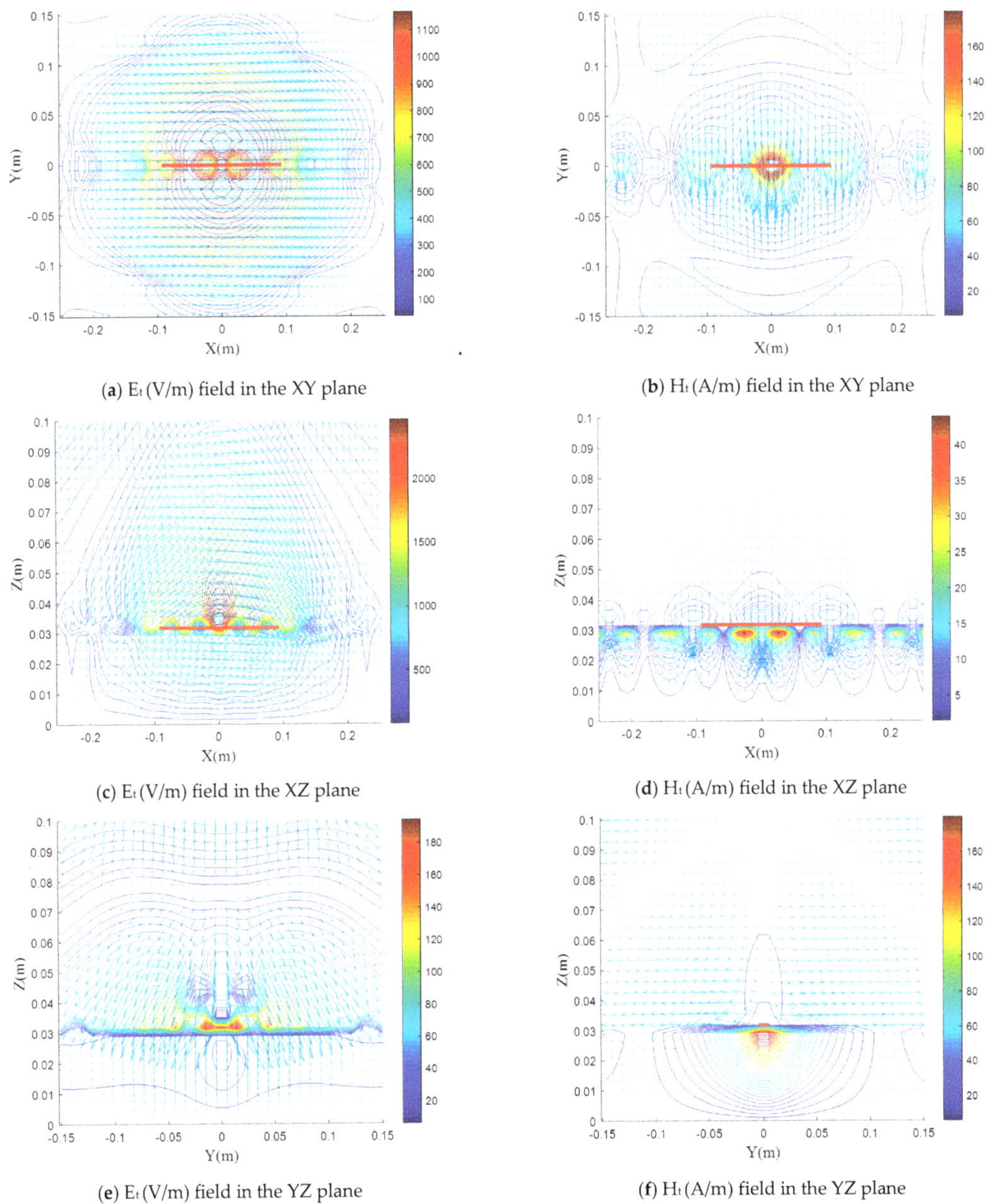

(**a**) E_t (V/m) field in the XY plane

(**b**) H_t (A/m) field in the XY plane

(**c**) E_t (V/m) field in the XZ plane

(**d**) H_t (A/m) field in the XZ plane

(**e**) E_t (V/m) field in the YZ plane

(**f**) H_t (A/m) field in the YZ plane

Figure 3. (**a–f**): Fields distribution plots in the transverse plane for the isotropic case ($\varepsilon_t = \varepsilon_z = 3.25$ and $\mu_t = \mu_z = 1$).

From the plots, we can easily notice that the E- and H-plane are the XZ and the YZ planes (Figure 3b,c), respectively. In Figure 3a, a quite rectangular shape of the E field distribution is clearly observed due to the fringing fields close to the edge of the strip and the electric field has a small component parallel to the dipole. The electric field is not entirely confined within the dielectric substrate; it extends partially into the air above. This phenomenon is commonly referred to as fringing, as shown in Figure 3a,c,e.

The fringing E-field makes the dipole electrically longer than its real physical dimension. This is why the dipole physical length is chosen to be slightly less than half the dielectric wavelength. At resonance, the dipole electrical length must equal half the dielectric wavelength. The antenna fringing field amount is strongly related to the radiating element geometry and the substrate permittivity. Substrates with low dielectric constants allow more field fringing than do high dielectric constant-substrates [45]. It can be seen from Figure 3e, given that the direction of propagation is along z-axis, that the spacing between the lines at z = 0.75 m is equivalent to $\lambda/4$, this illustrates the concept of propagation.

Figure 3d,f show a high confinement of the H field in the substrate, due to the fact that the magnetic field is in direct relation with the electric induction, this is due to the contribution of the two dielectric constants $\varepsilon_r = 3.25$ and $\mu_r = 1$.

5.3. Effect of the Electrical Uniaxial Anisotropy on the Electromagnetic-Field Distributions

In this section, we examine the effect of the electrical anisotropy on the electromagnetic field distributions in the three planes, and we interpret these results through the shape of the input impedance. Figure 4a–f show the normalized tangential electric field magnitude in a color plot superimposed on the vector plots in arrow representation for various values of ε_z with $\varepsilon_t = 3.25$, $\mu_z = 1$ and $\mu_t = 1$ in the XY, XZ, and YZ planes.

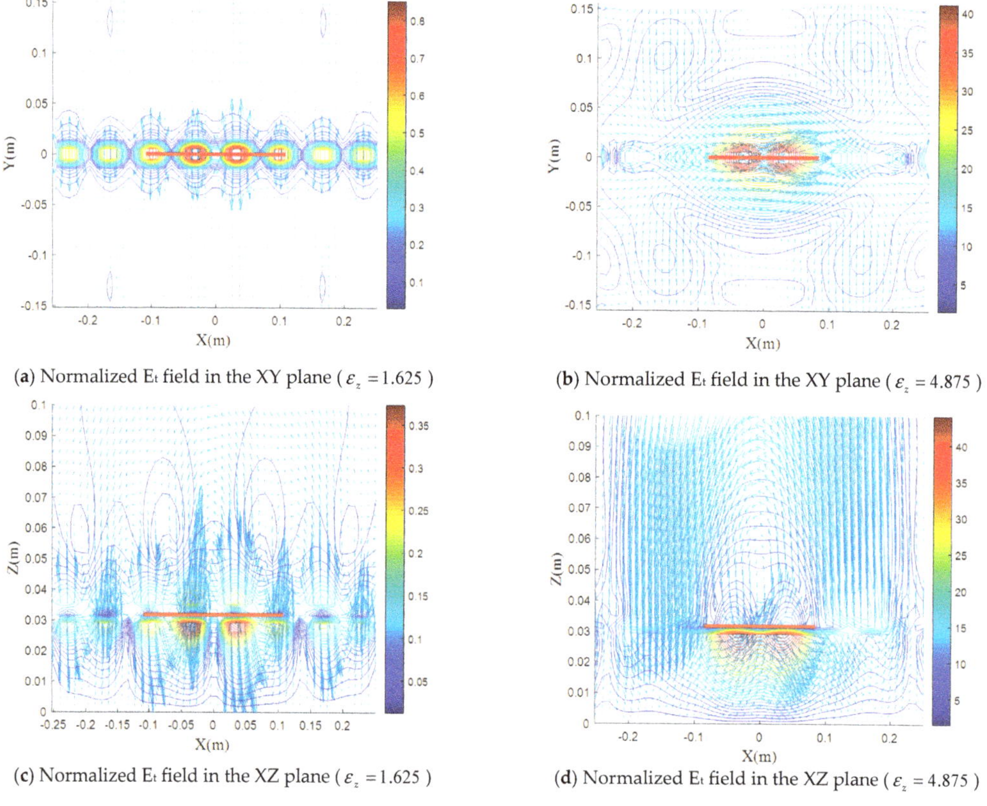

(**a**) Normalized E$_t$ field in the XY plane ($\varepsilon_z = 1.625$)

(**b**) Normalized E$_t$ field in the XY plane ($\varepsilon_z = 4.875$)

(**c**) Normalized E$_t$ field in the XZ plane ($\varepsilon_z = 1.625$)

(**d**) Normalized E$_t$ field in the XZ plane ($\varepsilon_z = 4.875$)

Figure 4. *Cont.*

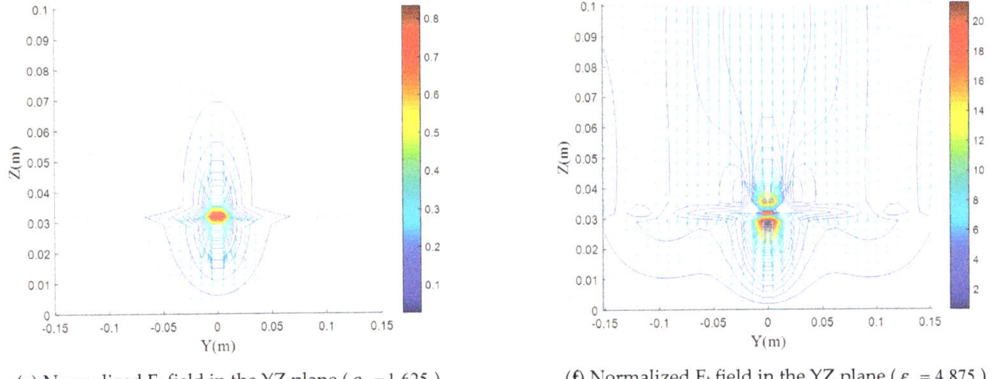

(**e**) Normalized E_t field in the YZ plane ($\varepsilon_z = 1.625$)   (**f**) Normalized E_t field in the YZ plane ($\varepsilon_z = 4.875$)

Figure 4. (**a–f**) Normalized electric field distributions for various values of ε_z with $\varepsilon_t = 3.25$, $\mu_z = 1$ and $\mu_t = 1$ in the XY, XZ, and YZ planes.

According to Figures 4 and 5, we roughly notice an effect reversed in shape and different in details of the two anisotropic elements ε_z and ε_t. A decrease in ε_z up to 40% leads to a slight decrease and the E-plane decreases its max by 50%, while an increase of 40% reveals a significant increase in the electric field in the E- and H-plane.

A decrease of 40% in ε_t leads to an almost identical increase (3 times) of the electric field maximum, while for an increase of 40%, an identical decrease of the maximum in all three planes is noticed (64%). As a consequence of this result, and to improve the dipole radiation, it is necessary to strongly minimize ε_t and increase, in the possibility, ε_z. From Figure 4a,b and Figure 5a,b, it can be seen that for the case ($\varepsilon_t = 1.625$), a maximum electric field exceeds 0.2 m, when the element ε_t is reduced, this agrees well with the literature [45].

According to Figure 4c,d and Figure 5c,d, the electric field lines are condensed since they are presented, in this case, in the E plane, and we can see that the component Ex is the most important in the cases of Figures 4d and 5c from Figure 4e,f and Figure 5e,f, it is clear that the most important part of the electric field is confined in the dielectric and around the dipole.

From Figures 6 and 7, we notice that the contribution of the two electrical anisotropic components ε_z and ε_t on the magnetic field distribution is almost the same as that on the electric field, except that the magnetic field in the E plane undergoes more decrease in the case (XZ plane $\varepsilon_z = 1.625$ (Figure 6c) and XZ plane $\varepsilon_t = 1.625$ (Figure 7c) compared to the other planes) and a slight increase for the case XZ plane $\varepsilon_z = 4.875$ (Figure 6d) compared to the other planes) and for the case $\varepsilon_t = 4.875$, the decrease is almost the same (Figure 7b,f).

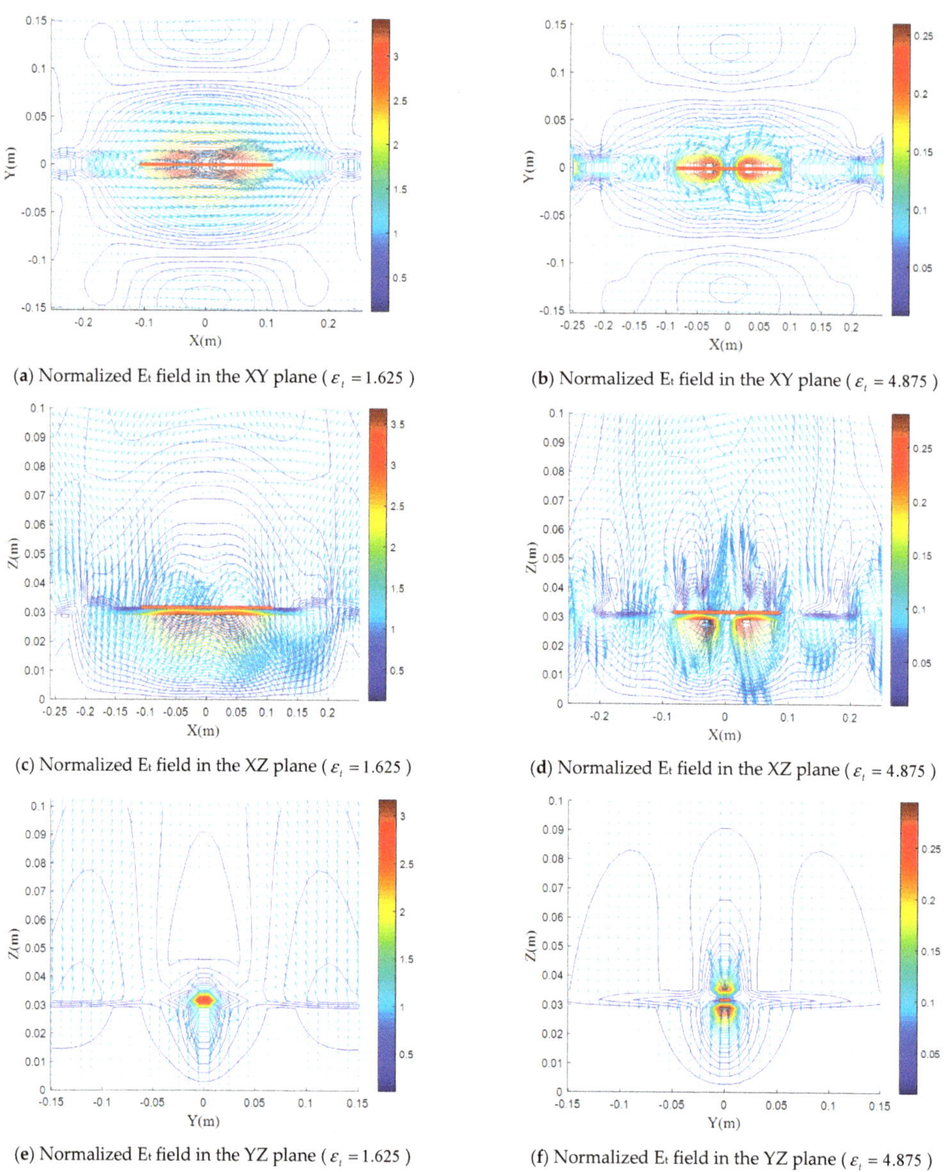

(**a**) Normalized E_t field in the XY plane ($\varepsilon_t = 1.625$)

(**b**) Normalized E_t field in the XY plane ($\varepsilon_t = 4.875$)

(**c**) Normalized E_t field in the XZ plane ($\varepsilon_t = 1.625$)

(**d**) Normalized E_t field in the XZ plane ($\varepsilon_t = 4.875$)

(**e**) Normalized E_t field in the YZ plane ($\varepsilon_t = 1.625$)

(**f**) Normalized E_t field in the YZ plane ($\varepsilon_t = 4.875$)

Figure 5. (**a–f**) Normalized electric field distributions for various values of ε_t with $\varepsilon_z = 3.25$, $\mu_z = 1$ and $\mu_t = 1$ in the XY, XZ, and YZ planes.

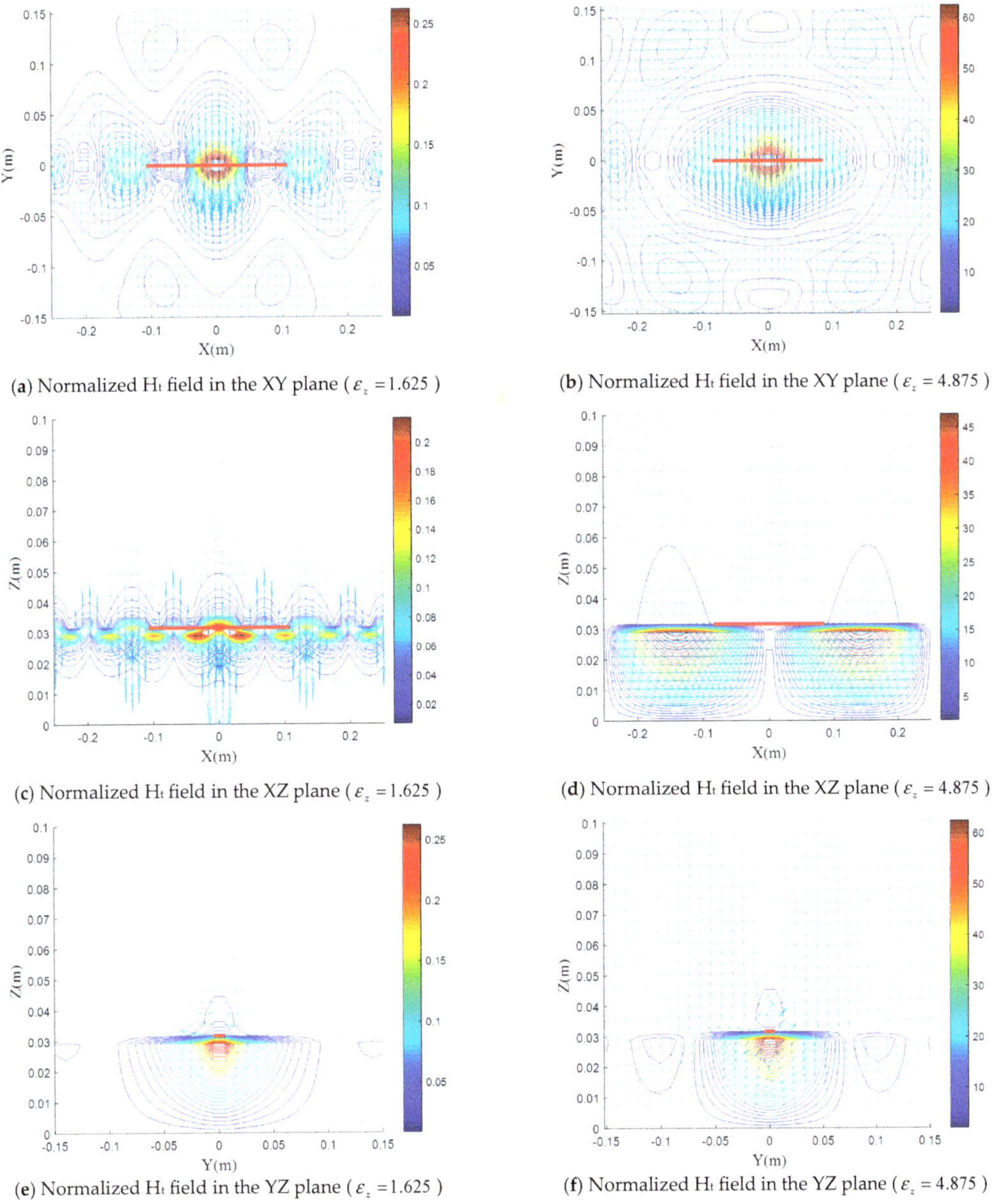

(**a**) Normalized H_t field in the XY plane ($\varepsilon_z = 1.625$)

(**b**) Normalized H_t field in the XY plane ($\varepsilon_z = 4.875$)

(**c**) Normalized H_t field in the XZ plane ($\varepsilon_z = 1.625$)

(**d**) Normalized H_t field in the XZ plane ($\varepsilon_z = 4.875$)

(**e**) Normalized H_t field in the YZ plane ($\varepsilon_z = 1.625$)

(**f**) Normalized H_t field in the YZ plane ($\varepsilon_z = 4.875$)

Figure 6. (**a–f**) Normalized magnetic field distributions for various values of ε_z with $\varepsilon_t = 3.25$, $\mu_z = 1$ and $\mu_t = 1$ in the XY, XZ and YZ planes.

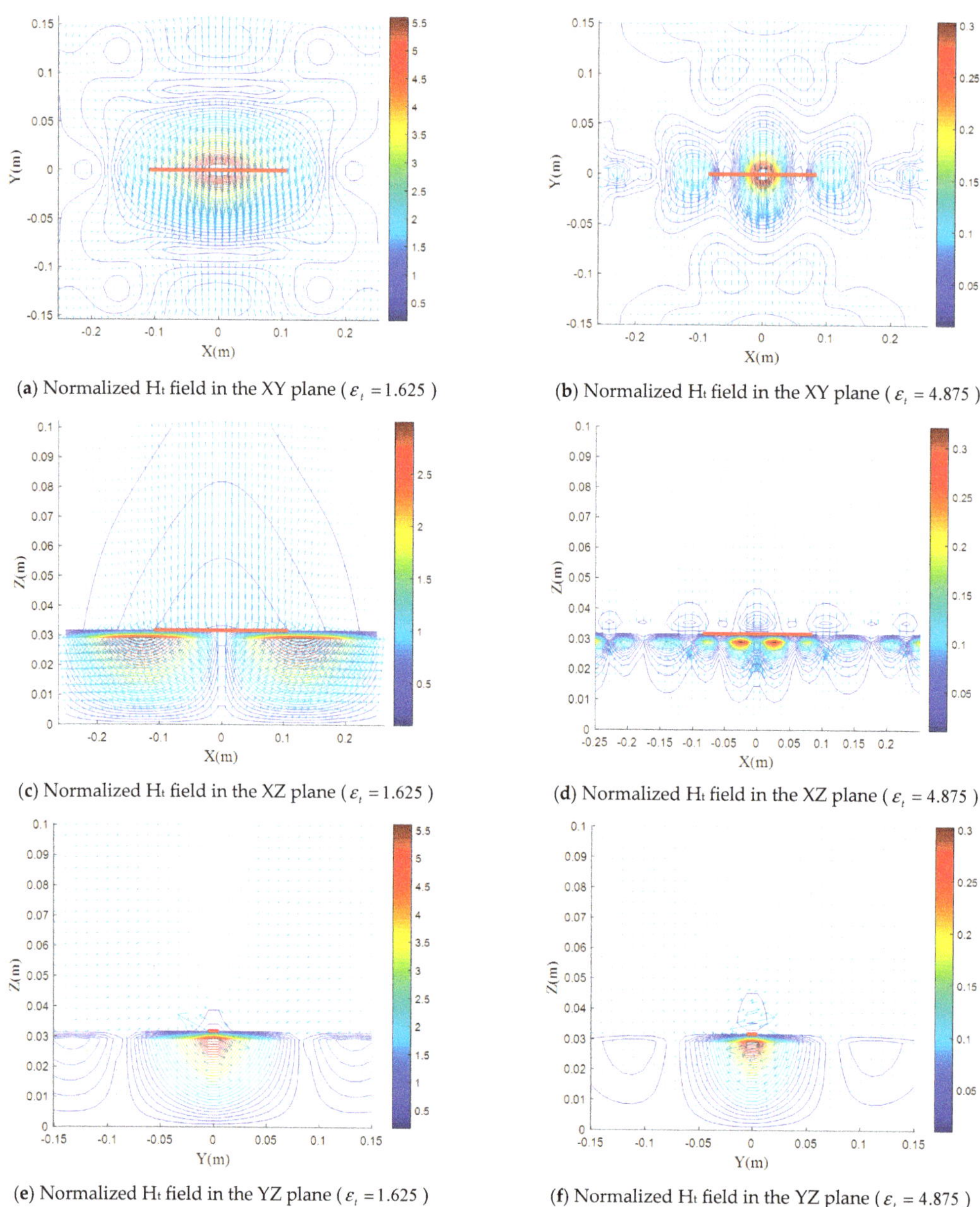

(**a**) Normalized H$_t$ field in the XY plane ($\varepsilon_t = 1.625$)

(**b**) Normalized H$_t$ field in the XY plane ($\varepsilon_t = 4.875$)

(**c**) Normalized H$_t$ field in the XZ plane ($\varepsilon_t = 1.625$)

(**d**) Normalized H$_t$ field in the XZ plane ($\varepsilon_t = 4.875$)

(**e**) Normalized H$_t$ field in the YZ plane ($\varepsilon_t = 1.625$)

(**f**) Normalized H$_t$ field in the YZ plane ($\varepsilon_t = 4.875$)

Figure 7. (**a–f**) Normalized magnetic field distributions for various values of ε_t with $\varepsilon_z = 3.25$, $\mu_z = 1$ and $\mu_t = 1$ in the XY, XZ, and YZ planes.

5.4. Effect of the Magnetic Uniaxial Anisotropy on Electromagnetic-Field Distributions

Figures 8–11 show the effect of the uniaxial electrical anisotropy on electric and magnetic field distributions.

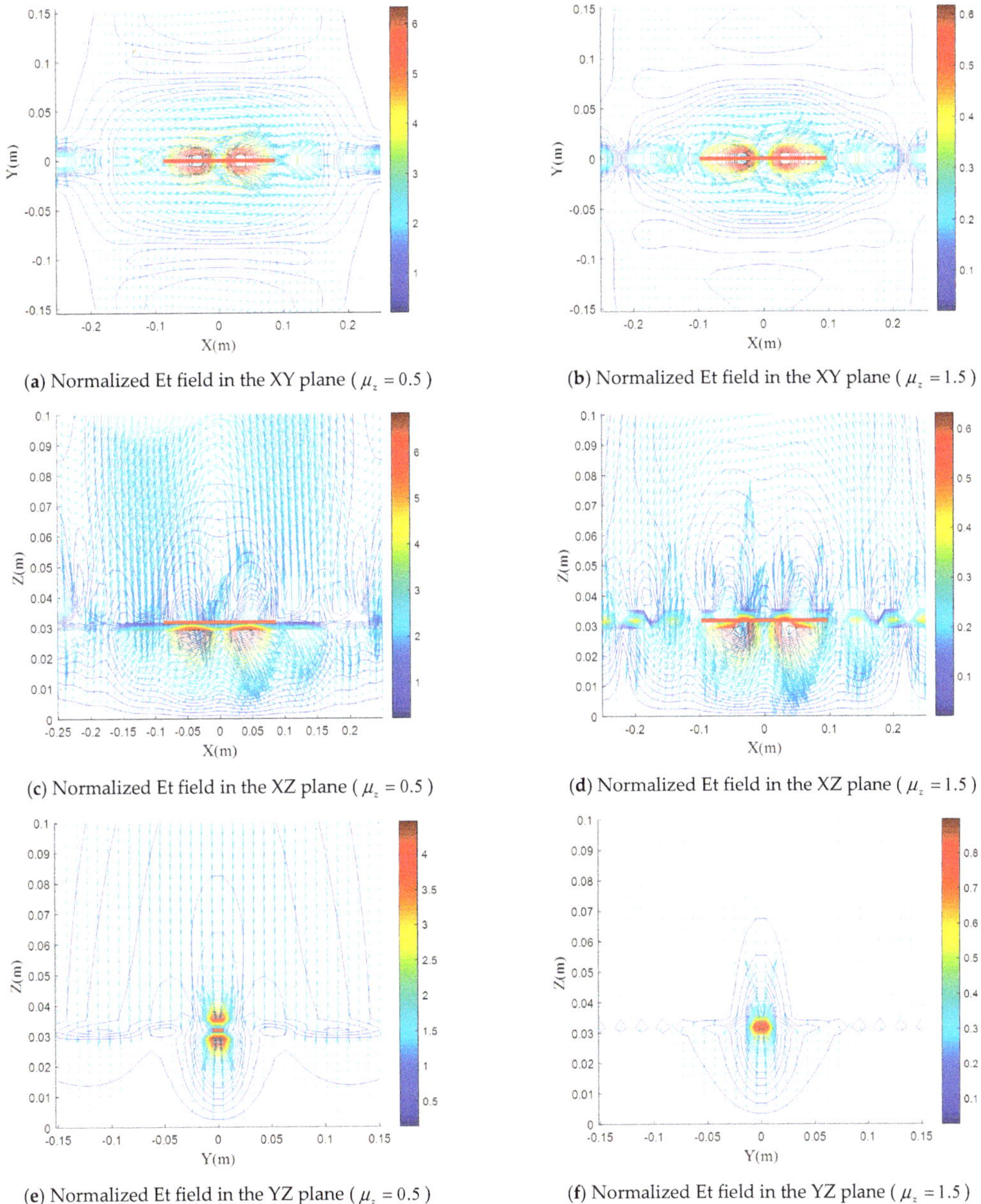

(**a**) Normalized Et field in the XY plane ($\mu_z = 0.5$)

(**b**) Normalized Et field in the XY plane ($\mu_z = 1.5$)

(**c**) Normalized Et field in the XZ plane ($\mu_z = 0.5$)

(**d**) Normalized Et field in the XZ plane ($\mu_z = 1.5$)

(**e**) Normalized Et field in the YZ plane ($\mu_z = 0.5$)

(**f**) Normalized Et field in the YZ plane ($\mu_z = 1.5$)

Figure 8. (**a**–**f**) Normalized electric field distributions for various values of μ_z with $\varepsilon_z = \varepsilon_t = 3.25$ and $\mu_t = 1$, in the XY, XZ, and YZ planes.

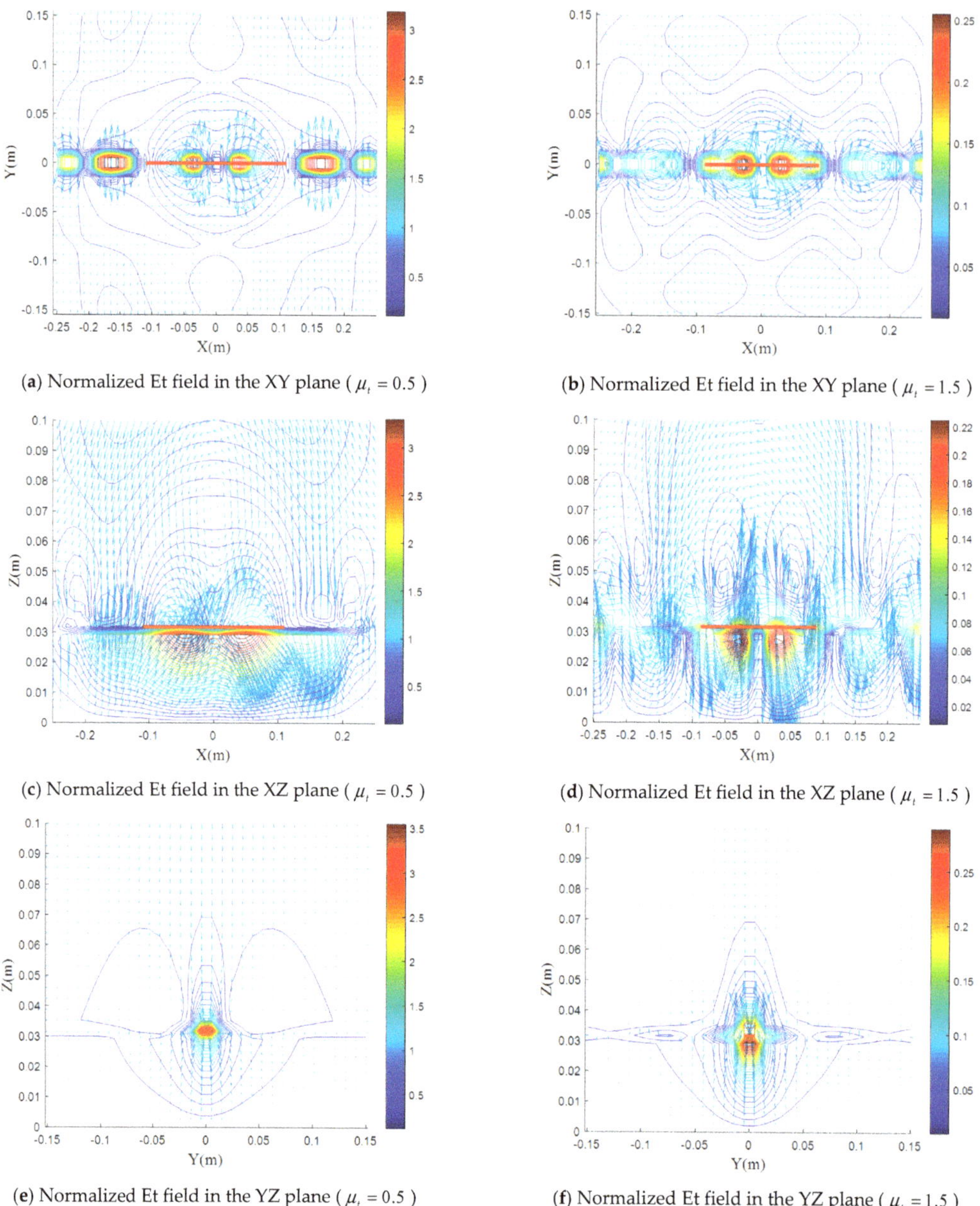

(**a**) Normalized Et field in the XY plane ($\mu_t = 0.5$)

(**b**) Normalized Et field in the XY plane ($\mu_t = 1.5$)

(**c**) Normalized Et field in the XZ plane ($\mu_t = 0.5$)

(**d**) Normalized Et field in the XZ plane ($\mu_t = 1.5$)

(**e**) Normalized Et field in the YZ plane ($\mu_t = 0.5$)

(**f**) Normalized Et field in the YZ plane ($\mu_t = 1.5$)

Figure 9. (**a–f**) Normalized electric field distributions for various values of μ_t with $\varepsilon_z = \varepsilon_t = 3.25$ and $\mu_z = 1$, in the XY, XZ, and YZ planes.

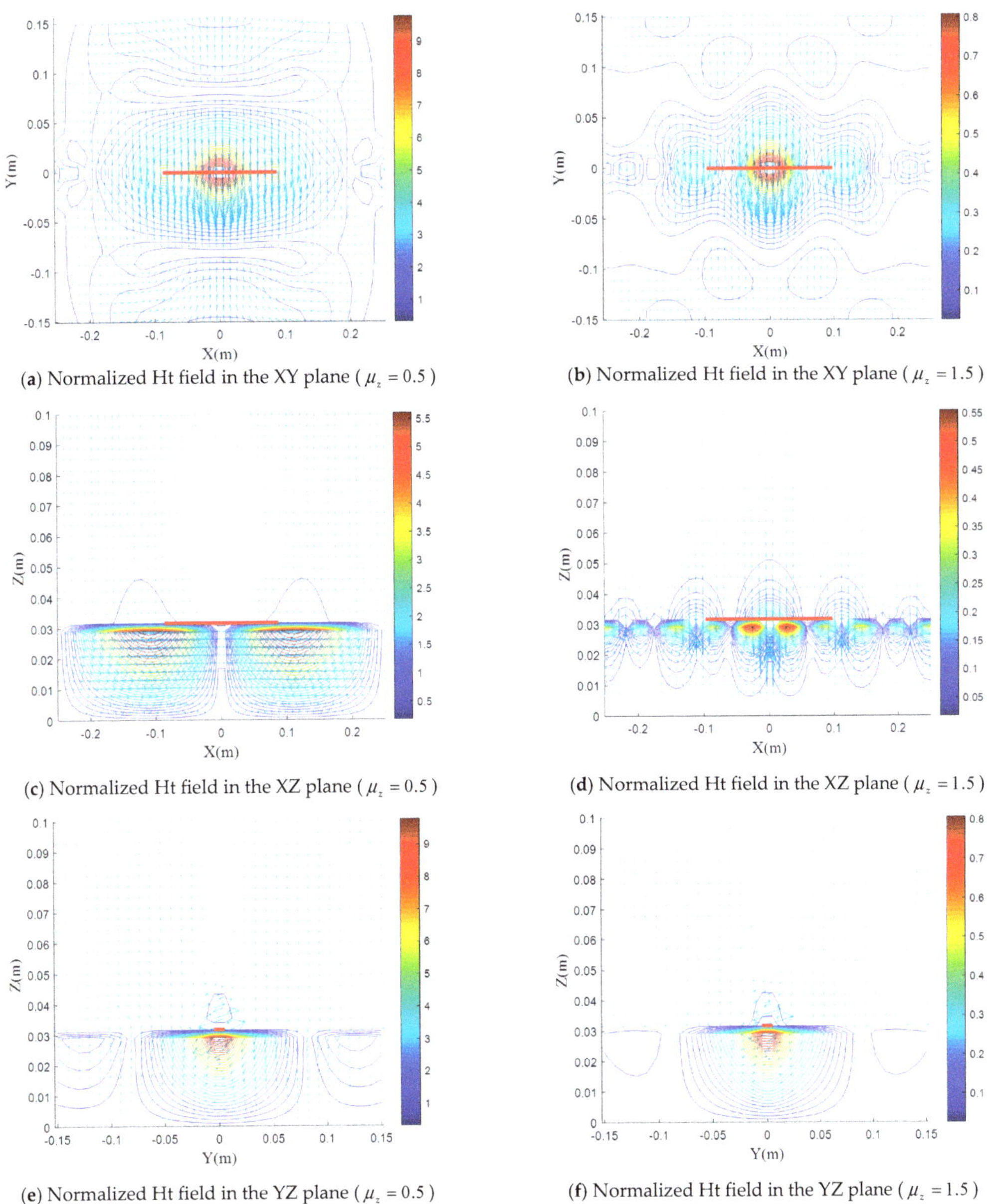

(**a**) Normalized Ht field in the XY plane ($\mu_z = 0.5$)

(**b**) Normalized Ht field in the XY plane ($\mu_z = 1.5$)

(**c**) Normalized Ht field in the XZ plane ($\mu_z = 0.5$)

(**d**) Normalized Ht field in the XZ plane ($\mu_z = 1.5$)

(**e**) Normalized Ht field in the YZ plane ($\mu_z = 0.5$)

(**f**) Normalized Ht field in the YZ plane ($\mu_z = 1.5$)

Figure 10. (**a**–**f**) Normalized magnetic field distributions for various values of μ_z with $\varepsilon_z = \varepsilon_t = 3.25$ and $\mu_t = 1$, in the XY, XZ, and YZ plane.

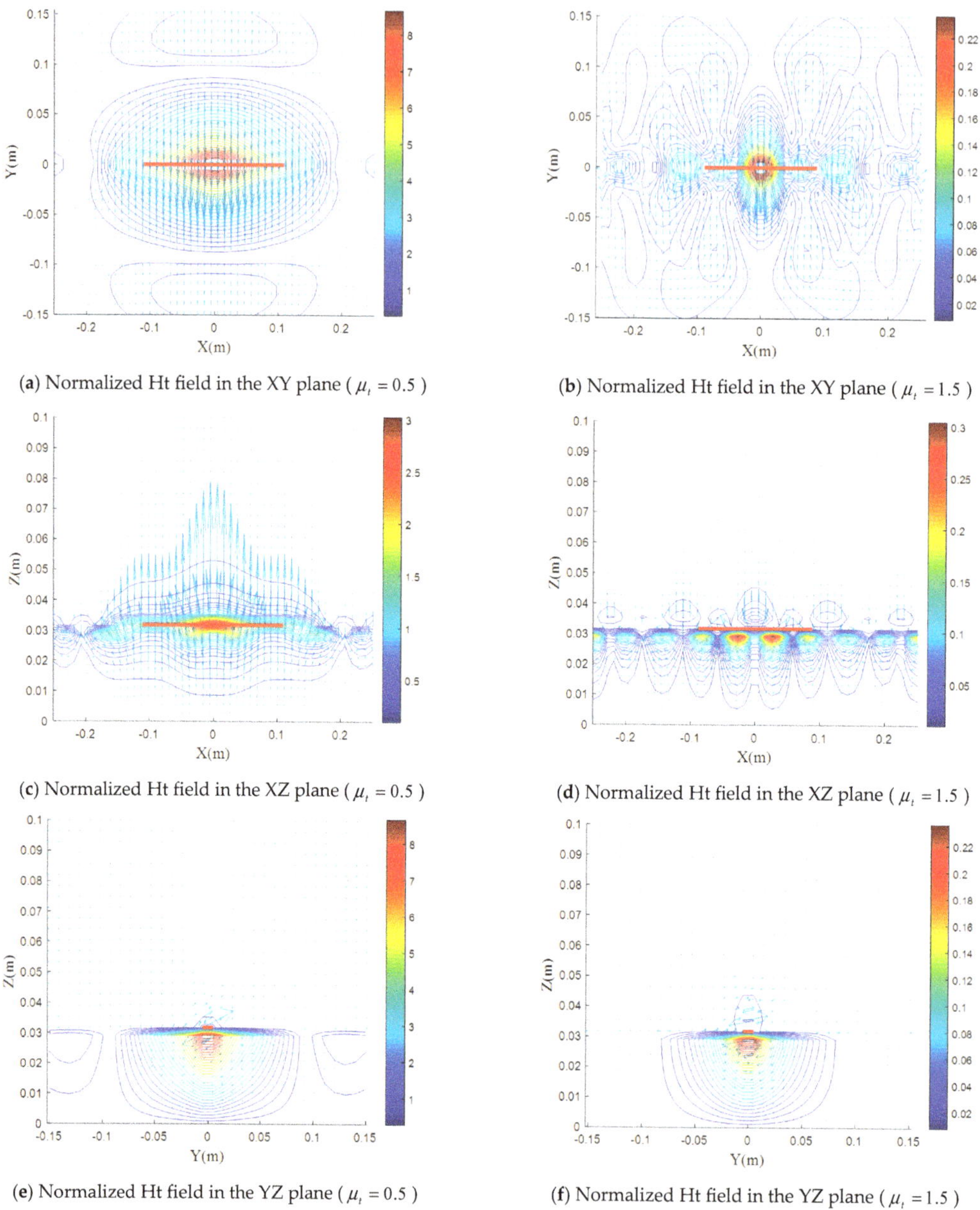

Figure 11. (**a–f**) Normalized magnetic field distributions for various values of μ_t with $\varepsilon_z = \varepsilon_t = 3.25$ and $\mu_z = 1$, in the XY, XZ, and YZ plane.

It can be seen from Figures 4, 5 and 8 that the effect of the permeability component μ_z on the line shape and distribution of the electric field is similar to that of ε_z, but in terms of value and quantity, it has the same effect as ε_t. However in the case of the μ_t component, and according to Figures 4, 5 and 9, the effect of μ_t component is close in shape and quantity to that of ε_t.

From here, we can note that the four (ε_z, ε_t, μ_t and μ_z) constitutive components have a different contribution to each other, and this is mainly due to the dipole antenna configurations and its positioning.

From Figures 8–11, we can see that the two components μ_t and μ_z have a different effect than the two permittivity components.

The increase of μ_t and μ_z leads to a decrease of the fields maximum compared to the isotropic case, while when ε_z increases, the maximum E increases strongly. Moreover, the effect of the two components μ_t and μ_z is close to that of ε_t in form and magnitude. This is due to the lines of the magnetic field, which rotates around the dipole (YZ plane), indicating that the maximum magnetic field interaction is through the two components μ_t and μ_z as well as with ε_t.

5.5. Effect of the Uniaxial Anisotropy on Input Impedance

Figure 12a shows the effect of ε_z on the input impedance. It consists in shifting the resonant length of the dipole antenna with a slight change in its peak, while ε_t affects significantly the magnitude of the input impedance with an increase and decrease of its peak, from 3 kΩ for $\varepsilon_t = 3.25$ to 5 kΩ for $\varepsilon_t = 1.625$ and 3 kΩ for $\varepsilon_t = 3.25$ to 2.5 kΩ for $\varepsilon_t = 4.875$, all with a slight shift in the resonant length (Figure 12b).

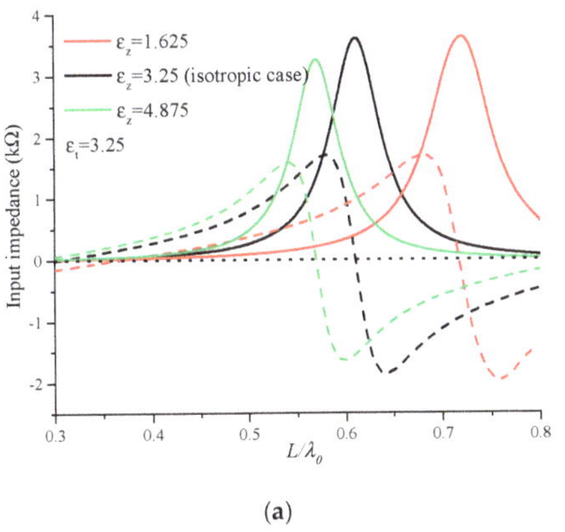

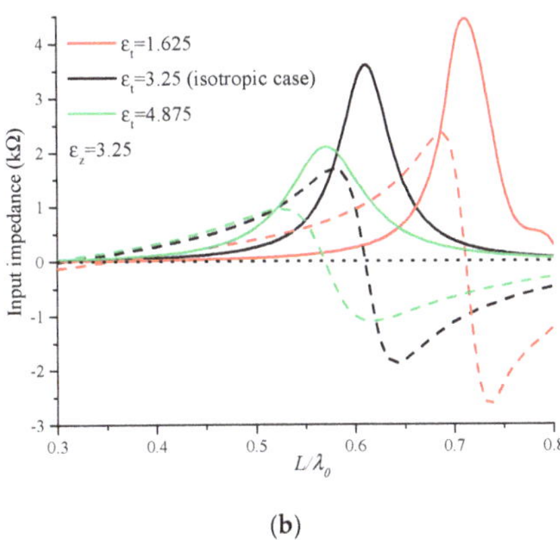

Figure 12. Real and imaginary parts of the input impedance for various values of (**a**): ε_z and (**b**): ε_t.

In Figure 13a,b, the effect of the two components of permeability μ_z and μ_t does not resemble that of the permittivity components ε_z and ε_t. An increase in μ_z results in an increase in the input impedance peak, with a decrease in the resonance frequency. The effect of the permeability component μ_t is reversed in this case, where an increase in the μ_t component leads to a significant increase in the resonance frequency with a decrease in the peak value of input impedance.

In the case $\varepsilon_z = 1.625$, a decrease of 35% of the maximum of the electric field is observed in the XZ plane (E plane), according to Figure 4c, accompanied by a close decrease in the magnetic field (27%) in the YZ plane (H plane), according to Figure 6e. This is translated by a slight increase (5.5 and 3.5 times) of the input impedance (Figure 12a). However, in the case for $\varepsilon_t = 1.625$, the maximum of the magnetic field in the H-plane (Figure 7e) is almost twice that of the electric field in the E-plane (Figure 5c) and for $\varepsilon_t = 4.875$, a ratio between the field maxima is 1.4. This illustrates the effect of ε_t on the input impedance (Figure 12b). In this case, the maximum of the magnetic field is greater than that of the electric field; this reveals the effect of the component ε_t on input impedance.

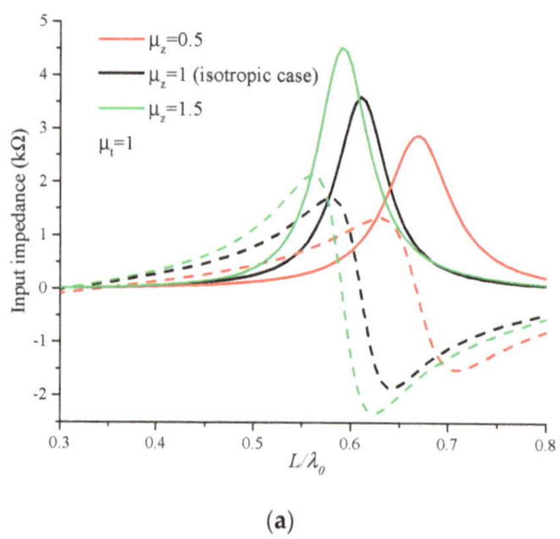
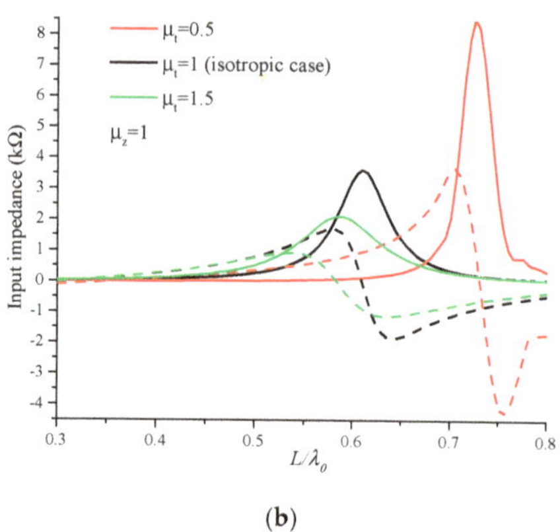

(a)   (b)

Figure 13. Real and imaginary parts of the input impedance for various values of (**a**): μ_z and (**b**): μ_z.

Figure 13a,b show the effect of the uniaxial magnetic anisotropy on the dipole input impedance. For the cases $\mu_z = 0.5$ and $\mu_z = 1.5$, the same ratio of increase and decrease between the electric field E and the magnetic field H (normalized values 7 times to 9 times (Figure 10), 0.8 to 0.65 (Figure 8), respectively) led to an almost identical decrease and increase, respectively, in input impedance (Figure 13a).

In the case $\mu_t = 0.5$, a 9-times increase of the H field maximum, in the YZ plane (Figure 11e), compared to the isotropic case, is accompanied by a 6-times increase of the E field maximum in the XZ plane (Figure 9c); this is translated by a strong increase of the input impedance (Figure 13b). On the other hand, in the case $\mu_t = 1.5$, an almost identical decrease of the H- and E-field maxima (0.25 (Figure 11f) and 0.22 (Figure 9d), respectively, in the YZ XZ plane); this is translated by a decrease of input impedance (Figure 13b).

To summarize, we notice that the choice of the optical axis of the uniaxial anisotropy and the geometry of the printed dipole lead to an asymmetrical Green's tensor, relating the electric field to the current density through the four constituent parameters ε_t, ε_z, μ_t and μ_z. This is the factor behind the difference registered between the effects of these components, in addition to the electromagnetic field distributions.

6. Conclusions

In this paper, we presented a theoretical study for the investigation of the electromagnetic field distributions and the input impedance of a printed dipole antenna structure loaded on a uniaxial anisotropic medium. The presentation of the electromagnetic field distributions, for which some examples have been shown here, provides a better understanding of the constitutive parameters (ε_t, ε_z, μ_t and μ_z) contributions. Furthermore, the electrical and magnetic uniaxial anisotropy offers more degrees of freedom and further flexibility to realize a good direct matching effect on the input impedance. This show that the complex media present a great potential in the design of innovative microwave components. It constitutes a starting point for further works to a better understanding of the behavior of the electromagnetic field in anisotropic and bianisotropic media and many more interesting results are expected.

Author Contributions: Design and concept, M.L.B., D.S., and C.Z.; methodology, C.Z., I.E., and M.A.; investigation, M.A. and F.F.; resources, C.Z. and I.E.; writing—original draft preparation, M.L.B., C.Z., and D.S.; writing—review and editing, I.E., R.A.A.-A., and M.A.; supervision, R.A.A.-A. and E.L.; project administration, J.R. and E.L., formal analysis, M.L.B. and C.Z. All authors have read and agreed to the published version of the manuscript.

Funding: This project received funding in part from the DGRSDT (Direction Générale de la Recherche Scientifique et du Développement Technologique), MESRS (Ministry of Higher Education and Scientific Research), Algeria. This work is also funded by the FCT/MEC through national funds and when applicable co-financed by the ERDF, under the PT2020 Partnership Agreement, the UID/EEA/50008/2019 project.

Data Availability Statement: All data are included within manuscript.

Acknowledgments: This work was supported in part by the DGRSDT (Direction Générale de la Recherche Scientifique et du Développement Technologique), MESRS (Ministry of Higher Education and Scientific Research), Algeria. This work is part of the POSITION-II project funded by the ECSEL joint Undertaking under grant number Ecsel-7831132-Postitio-II-2017-IA. This work is partially supported by RTI2018-095499-B-C31, funded by Ministerio de Ciencia, Innovación y Universidades, Gobierno de España (MCIU/AEI/FEDER, UE).

Conflicts of Interest: The authors declare no conflict of interest.

References

1. Kirilenko, A.A.; Steshenko, S.O.; Derkach, V.N.; Prikolotin, S.A.; Kulik, D.Y.; Prosvirnin, S.; Mospan, L.P. Rotation of the polarization plane by double-layer planar-chiral structures. Review of the results of theoretical and experimental studies. *Radioelectron. Commun. Syst.* **2017**, *60*, 193–205. [CrossRef]
2. Akdagli, A. *Behaviour of Electromagnetic Waves in Different Media and Structures*, 1st ed.; BoD–Books on Demand: Rijeka, Croatia, 2011.
3. Guo, B. Photonic band gap structures of obliquely incident electromagnetic wave propagation in a one-dimension absorptive plasma photonic crystal. *Phys. Plasmas* **2009**, *16*, 043508. [CrossRef]
4. Krowne, C.M. Left-handed material anisotropy effect on guided wave electromagnetic fields. *J. Appl. Phys.* **2006**, *99*, 044914. [CrossRef]
5. Krowne, C.M.; Daniel, M. Electromagnetic field behavior in dispersive isotropic negative phase velocity/negative refractive index guided wave structures compatible with millimeter-wave monolithic integrated circuits. *J. Nanomater.* **2007**, *2007*, 054568. [CrossRef]
6. Krowne, C.M. Electromagnetic-field theory and numerically generated results for propagation in left-handed guided-wave single-microstrip structures. *IEEE Trans. Microw. Theory Tech.* **2003**, *51*, 2269–2283. [CrossRef]
7. Krowne, C.M. Electromagnetic distributions demonstrating asymmetry using a spectral-domain dyadic Green's function for ferrite microstrip guided-wave structures. *IEEE Trans. Microw. Theory Tech.* **2005**, *53*, 1345–1361. [CrossRef]
8. Zebiri, C.; Lashab, M.; Benabdelaziz, F. Effect of anisotropic magneto-chirality on the characteristics of a microstrip resonator. *IET Microw. Antennas Propag.* **2010**, *4*, 446–452. [CrossRef]
9. Zebiri, C.; Lashab, M.; Benabdelaziz, F. Rectangular microstrip antenna with uniaxial bi-anisotropic chiral substrate–superstrate. *IET Microw. Antennas Propag.* **2011**, *5*, 17–29. [CrossRef]
10. Sayad, D.; Zebiri, C.; Daoudi, S.; Benabdelaziz, F. Analysis of the effect of a gyrotropic anisotropy on the phase constant and characteristic impedance of a shielded microstrip line. *Adv. Electromagn.* **2019**, *8*, 15–22. [CrossRef]
11. Heydari, M.B.; Ahmadvand, A. A novel analytical model for a circularly-polarized; ferrite-based slot antenna by solving an integral equation for the electric field on the circular slot. *Waves Random Complex Media* **2020**, 1–20. [CrossRef]
12. Lee, W.; Hong, Y.K.; Choi, M.; Won, H.; Lee, J.; Park, S.O.; Yoon, H.S. Ferrite-cored patch antenna with suppressed harmonic radiation. *IEEE Trans. Antennas Propag.* **2018**, *66*, 3154–3159. [CrossRef]
13. Kamra, V.; Dreher, A. Efficient analysis of multiple microstrip transmission lines with anisotropic substrates. *IEEE Microw. Wirel. Compon. Lett.* **2018**, *28*, 636–638. [CrossRef]
14. Buzov, A.L.; Buzova, M.A.; Klyuev, D.S.; Mishin, D.V.; Neshcheret, A.M. Calculating the Input Impedance of a Microstrip Antenna with a Substrate of a Chiral Metamaterial. *J. Commun. Technol. Electron.* **2018**, *63*, 1259–1264. [CrossRef]
15. Klyuev, D.S.; Minkin, M.A.; Mishin, D.V.; Neshcheret, A.M.; Tabakov, D.P. Characteristics of Radiation from a Microstrip Antenna on a Substrate Made of a Chiral Metamaterial. *Radiophys. Quantum Electron.* **2018**, *61*, 445–455. [CrossRef]
16. Hu, Y.; Fang, Y.; Wang, D.; Zhan, Q.; Zhang, R.; Liu, Q.H. The scattering of electromagnetic fields from anisotropic objects embedded in anisotropic multilayers. *IEEE Trans. Antennas Propag.* **2019**, *67*, 7561–7568. [CrossRef]
17. Zebiri, C.; Daoudi, S.; Benabdelaziz, F.; Lashab, M.; Sayad, D.; Ali, N.T.; Abd-Alhameed, R.A. Gyro-chirality effect of bianisotropic substrate on the operational of rectangular microstrip patch antenna. *Int. J. Appl. Electromagn. Mech.* **2016**, *51*, 249–260. [CrossRef]
18. Zebiri, C.; Benabdelaziz, F.; Sayad, D. Surface waves investigation of a bianisotropic chiral substrate resonator. *Prog. Electromagn. Res.* **2012**, *40*, 399–414. [CrossRef]
19. Eroglu, A.; Lee, J.K. Far field radiation from an arbitrarily oriented Hertzian dipole in the presence of a layered anisotropic medium. *IEEE Trans. Antennas Propag.* **2005**, *53*, 3963–3973. [CrossRef]
20. Sayad, D.; Benabdelaziz, F.; Zebiri, C.; Daoudi, S.; Abd-Alhameed, R.A. Spectral domain analysis of gyrotropic anisotropy chiral effect on the input impedance of a printed dipole antenna. *Prog. Electromagn. Res.* **2016**, *51*, 1–8. [CrossRef]
21. Braaten, B.D.; Rogers, D.A.; Nelson, R.M. Multi-conductor spectral domain analysis of the mutual coupling between printed dipoles embedded in stratified uniaxial anisotropic dielectrics. *IEEE Trans. Antennas Propag.* **2012**, *60*, 1886–1898. [CrossRef]

22. Soares, A.; Fonseca, S.B.D.A.; Giarola, A. The effect of a dielectric cover on the current distribution and input impedance of printed dipoles. *IEEE Trans. Antennas Propag.* **1984**, *32*, 1149–1153. [CrossRef]
23. Nelson, R.M.; Rogers, D.A.; D'Assuncao, A.G. Resonant frequency of a rectangular microstrip patch on several uniaxial substrates. *IEEE Trans. Antennas Propag.* **1990**, *38*, 973–981. [CrossRef]
24. Davidson, D.B.; Aberle, J.T. An introduction to spectral domain method-of-moments formulations. *IEEE Antennas Propag. Mag.* **2004**, *46*, 11–19. [CrossRef]
25. Wait, J. Fields of a horizontal dipole over a stratified anisotropic half-space. *IEEE Trans. Antennas Propag.* **1966**, *14*, 790–792. [CrossRef]
26. Kong, J.A. Electromagnetic fields due to dipole antennas over stratified anisotropic media. *Geophysics* **1972**, *37*, 985–996. [CrossRef]
27. Tang, C.M. Electromagnetic fields due to dipole antennas embedded in stratified anisotropic media. *IEEE Trans. Antennas Propag.* **1979**, *27*, 665–670. [CrossRef]
28. Lee, J.K.; Kong, J.A. Dyadic Green's functions for layered anisotropic medium. *Electromagnetics* **1983**, *3*, 111–130. [CrossRef]
29. Braaten, B.D.; Nelson, R.M.; Rogers, D.A. Input impedance and resonant frequency of a printed dipole with arbitrary length embedded in stratified uniaxial anisotropic dielectrics. *IEEE Antennas Wirel. Propag. Lett.* **2009**, *8*, 806–810. [CrossRef]
30. Eroglu, A.; Lee, Y.H.; Lee, J.K. Dyadic Green's functions for multi-layered uniaxially anisotropic media with arbitrarily oriented optic axes. *IET Microw. Antennas Propag.* **2011**, *5*, 1779–1788. [CrossRef]
31. Wang, N.; Wang, G.P. Effective medium theory with closed-form expressions for bi-anisotropic optical metamaterials. *Opt. Express* **2019**, *27*, 23739–23750. [CrossRef]
32. Erturk, V.B.; Rojas, R.G. Efficient analysis of input impedance and mutual coupling of microstrip antennas mounted on large coated cylinders. *IEEE Trans. Antennas Propag.* **2003**, *51*, 739–749. [CrossRef]
33. Ieukenov, S.K.; Assilbekova, A.M. Surface of wave vectors of electromagnetic waves in anisotropic dielectric media with rhombic symmetry. *Telecommun. Radio Eng.* **2017**, *76*, 1231–1238. [CrossRef]
34. Sayad, D.; Zebiri, C.; Elfergani, I.; Rodriguez, J.; Abobaker, H.; Ullah, A.; Benabdelaziz, F. Complex bianisotropy effect on the propagation constant of a shielded multilayered coplanar waveguide using improved full generalized exponential matrix technique. *Electronics* **2020**, *9*, 243. [CrossRef]
35. Sayad, D.; Zebiri, C.; Elfergani, I.; Rodriguez, J.; Abd-Alhameed, R.A.; Benabdelaziz, F. Analysis of Chiral and Achiral Medium Based Coplanar Waveguide Using Improved Full Generalized Exponential Matrix Technique. *Radioengineering* **2020**, *29*, 591–600. [CrossRef]
36. Zebiri, C.; Sayad, D. Effect of bianisotropy on the characteristic impedance of a shielded microstrip line for wideband impedance matching applications. *Waves Random Complex Media* **2020**, 1–14. [CrossRef]
37. Nakano, H.; Kerner, S.R.; Alexopoulos, N.G. The moment method solution for printed wire antennas of arbitrary configuration. *IEEE Trans. Antennas Propag.* **1988**, *36*, 1667–1674. [CrossRef]
38. Lee, H.; Tripathi, V.K. Spectral domain analysis of frequency dependent propagation characteristics of planar structures on uniaxial medium. *IEEE Trans. Microw. Theory Tech.* **1982**, *30*, 1188–1193.
39. Harrington, R.F. *Field Computation by Moment Methods*; IEEE Press, Inc.: New York, NY, USA, 1992.
40. Itoh, T. *Numerical Techniques for Microwave and Millimeter Wave Passive Structures*, 1st ed.; John Wiley and Sons: Hoboken, NJ, USA, 1988.
41. Bianconi, G.; Mittra, R. Efficient Numerical Techniques for Analyzing Microstrip Circuits and Antennas Etched on Layered Media via the Characteristic Basis Function Method. In *Computational Electromagnetics*; Springer: New York, NY, USA, 2014; pp. 111–148.
42. Rana, I.; Alexopoulos, N. Current distribution and input impedance of printed dipoles. *IEEE Trans. Antennas Propag.* **1981**, *29*, 99–105. [CrossRef]
43. Braaten, B.D.; Rogers, D.A.; Nelson, R.M. Current distribution of a printed dipole with arbitrary length embedded in layered uniaxial anisotropic dielectrics. In Proceedings of the 2009 SBMO/IEEE MTT-S International Microwave and Optoelectronics Conference (IMOC), Belem, Brazil, 3–6 November 2009; pp. 72–77.
44. *MATLAB*, version 2018a; The MathWorks, Inc.: Natick, MA, USA, 2018.
45. Codreanu, I.; Boreman, G.D. Influence of dielectric substrate on the responsivity of microstrip dipole-antenna-coupled infrared microbolometers. *Appl. Opt.* **2002**, *41*, 1835–1840. [CrossRef]

Article

8-Port Semi-Circular Arc MIMO Antenna with an Inverted L-Strip Loaded Connected Ground for UWB Applications

Tathababu Addepalli [1], Arpan Desai [2], Issa Elfergani [3,4,*], N. Anveshkumar [5], Jayshri Kulkarni [6], Chemseddine Zebiri [7], Jonathan Rodriguez [3] and Raed Abd-Alhameed [4,8]

[1] Department of ECE, JNTUA, Andhra Pradesh, Anantapur 515002, India; babu.478@gmail.com
[2] Department of Electronics and Communication Engineering, CSPIT, Charotar University of Science and Technology (CHARUSAT), Changa 388421, India; arpandesai.ec@charusat.ac.in
[3] Instituto de Telecomunicações, Campus Universitário de Santiago, 3810-193 Aveiro, Portugal; jonathan@av.it.pt
[4] School of Engineering and Informatics, University of Bradford, Bradford BD7 1DP, UK; r.a.a.abd@bradford.ac.uk
[5] Faculty of Electronics and Communications Engineering, VIT Bhopal University, Bhopal 466114, India; anvesh498@gmail.com
[6] Department of Electronics and Telecommunication Engineering, Vishwakarma Institute of Information Technology, Pune 411037, India; jayah2113@gmail.com
[7] Laboratoire D'Electronique de Puissance et Commande Industrielle (LEPCI), Department of Electronics, University of Ferhat Abbas, SETIF -1-, Setif 19000, Algeria; czebiri@univ-setif.dz
[8] Information and Comms Eng. Dept., Basrah University College of Science and Technology, Basrah 24001, Iraq
* Correspondence: i.t.e.elfergani@av.it.pt or i.elfergani@bradford.ac.uk; Tel.: +35-12-3437-7900

Citation: Addepalli, T.; Desai, A.; Elfergani, I.; Anveshkumar, N.; Kulkarni, J.; Zebiri, C.; Rodriguez, J.; Abd-Alhameed, R. 8-Port Semi-Circular Arc MIMO Antenna with an Inverted L-Strip Loaded Connected Ground for UWB Applications. *Electronics* **2021**, *10*, 1476. https://doi.org/10.3390/electronics10121476

Academic Editor: Athanasios Kanatas

Received: 19 May 2021
Accepted: 17 June 2021
Published: 19 June 2021

Publisher's Note: MDPI stays neutral with regard to jurisdictional claims in published maps and institutional affiliations.

Abstract: Multiple-input multiple-output (MIMO) antennas with four and eight elements having connected grounds are designed for ultra-wideband applications. Careful optimization of the lines connecting the grounds leads to reduced mutual coupling amongst the radiating patches. The proposed antenna has a modified substrate geometry and comprises a circular arc-shaped conductive element on the top with the modified ground plane geometry. Polarization diversity and isolation are achieved by replicating the elements orthogonally forming a plus shape antenna structure. The modified ground plane consists of an inverted L strip and semi ellipse slot over the partial ground that helps the antenna in achieving effective wide bandwidth spanning from (117.91%) 2.84–11 GHz. Both 4/8-port antenna achieves a size of $0.61\,\lambda \times 0.61\,\lambda\,mm^2$ (lowest frequency) where 4-port antenna is printed on FR4 substrate. The 4-port UWB MIMO antenna attains wide impedance bandwidth, Omni-directional pattern, isolation >15 dB, ECC < 0.015, and average gain >4.5 dB making the MIMO antenna suitable for portable UWB applications. Four element antenna structure is further extended to 8-element configuration with the connected ground where the decent value of IBW, isolation, and ECC is achieved.

Keywords: 4-elements; 8-elements; UWB; MIMO; ECC; isolation

1. Introduction

Wireless communications are getting popular at an astounding pace. The advancement in technology is helping the devices in shrinking their size day by day. Wireless communication when clubbed with various applications, such as medical, IoT, logistics, and personal area communication, can help various groups of people in many ways. The amalgamation can be very well-established with the deployment of WPANs that works on the IEEE protocols.

Increasing the bandwidth of such networks allows for a higher data rate. Due to the very broad frequency range spanning from 3.1 GHz to 10.6 GHz, ultra-wideband antennas are the most promising way to reach large bandwidth. Designing a UWB antenna is a difficult challenge since it must be capable of providing a big band following Federal Communications Commission (FCC) standards [1]. Multiple-input multiple-output (MIMO) technology is a key for enhancing the channel capacity by using multiple antennas and thus makes the antennas much more effective in densely populated areas by offering high throughput and data rates in the order of Gigabits/s [2]. The associated challenges of packing the elements closely while reducing the mutual coupling and achieving high isolation are of great importance in MIMO antennas.

Separate ground helps greatly in achieving the isolation however common ground is very much preferred to ensure the smooth operations of devices where antennas will be embedded [3].Various 4- and 8-element MIMO antennas are proposed in [4–30]. In [4–13], 4-port MIMO antennas with separate ground are proposed however connected ground configurations are considered for better commercial utilization of the MIMO antenna.

Researchers have proposed 4-elements connected ground UWB MIMO antennas in [14–21]. In [14], a 4-port CPW fed orthogonally arranged MIMO antenna is proposed having a bandwidth of 2.1–20 GHz however it has a notched band between 3.3–4.1 GHz and 8.2–8.6 GHz frequency bands. An antenna structure with four triangular-shaped monopole elements with a neutralization ring is proposed in [15] which spans from 138.21% (3.1–17.3 GHz). The monopole arrangement in orthogonal and symmetrical fashion leads to good isolation however the overall geometry gets complex due to the use of an additional neutralization ring. A 4-port UWB MIMO antenna having IBW of 130.43% (2.8–13.3 GHz) is proposed in [16] where rhombic-shaped identical monopoles are arranged orthogonally for achieving polarization diversity and better isolation. However, notch bands are introduced by using elliptical (CSRR) complementary split-ring resonator geometry. A reconfigurable MIMO antenna (1.77–2.51 GHz) integrated with sensing antenna 164.28% (0.75–7.65 GHz) for (CR) cognitive radio applications is proposed in [17]. The reconfigurability demands supply voltage which makes it less useful for portable applications. A compact UWB MIMO radiator having high isolation is presented in [18]. Stepped slots are utilized to realize the compact structure which spans from 115.78% (3.2–12 GHz). In [19], a common ground 4-port monopole MIMO antenna is proposed. Antenna achieves good MIMO performance however the bandwidth only spans from 58.63% (2.70–4.94 GHz). A 4-port MIMO antenna with reduced mutual coupling by introducing a neutralization line technique spanning from 96.47% (3.52–10.08 GHz) is proposed in [20]. The use of separate isolation techniques leads to design and fabrication complexities. Orthogonal diversity in a 4-port UWB antenna 119.14% (3.8–15 GHz) is achieved by aligning the monopole elements towards four sides of the substrate. The inclusion of ground stubs further helps in increasing the isolation between elements [21].

Researchers have proposed 4-port MIMO antennas using conformable substrates [22] and transparent materials [23,24] to make such antennas useful for their usage in locations where space is a constraint. However, flexible and transparent antennas face problems due to lower gain and efficiency.

The 8-port MIMO antennas may become impractical for their use in some scenarios due to the increase in size. In [28,30], the compactness of design exceeds 0.6 λ while the antennas proposed in [26,27,29,31] has a narrow bandwidth.

Therefore, in the proposed work, effective IBW of (117.91%) 2.84–11 GHz is achieved which is greater than work proposed in [12,13,19,20], no notch band unlike [14,16], no complex isolation mechanism [15,20,21], no active components [17], size of 0.61 λ × 0.61 λ mm² with satisfactory MIMO diversity parameter results thus making it commercially suitable for its use in portable applications.

In this work, a 4-port MIMO structure with microstrip feeding method is proposed for UWB applications. The microstrip feeding method is chosen in this work as it provides lower losses and ease of connector interfacing enabling effective use of low-cost FR-4 substrate. Further the antenna provides high isolation without using any complex isolation techniques in a connected ground configuration. The modified ground geometry helps in achieving wider bandwidth and more importantly enhances the antenna effectiveness.

Here, novelty lies in designing the main radiator of each antenna element which is a symmetrical ground-coupled antenna and is mainly composed of a circular arc-shaped monopole on the top side with modified ground structure coupled to the other grounds on the bottom side. Notably, the proposed 4-port MIMO antenna is realized by symmetrically arranging the four identical antenna elements in a sequential rotational manner for achieving good polarization diversity which is extended for 8-port MIMO antenna configuration. Other typical MIMO antenna performances, such as ECC and CCL, are also explicitly shown and validated by measuring the experimental values over the fabricated prototype.

The paper is arranged as follows: Proposed 2-port, 4-port, and 8-port antenna geometry are explained in Section 2 followed by results and discussion in Section 3, MIMO diversity parameters are explained in Section 4, time domain analysis in Sections 5 and 6 concludes the work.

2. Antenna Configuration and Parametric Study

The antenna geometry (top and bottom) of a single element that is a part of a 4-element antenna is visible in Figure 1a,b. A circular arc-shaped structure on the top and a modified ground plane on the backside is simulated over the FR4 (Flame Retardant-4) substrate. The modified ground plane consists of an inverted L strip and semi-ellipse slot over the partial ground. The perspective view of the antenna as shown in Figure 1c shows the conductive part (patch and ground) printed over the substrate of a thickness (T) 1.6 mm.

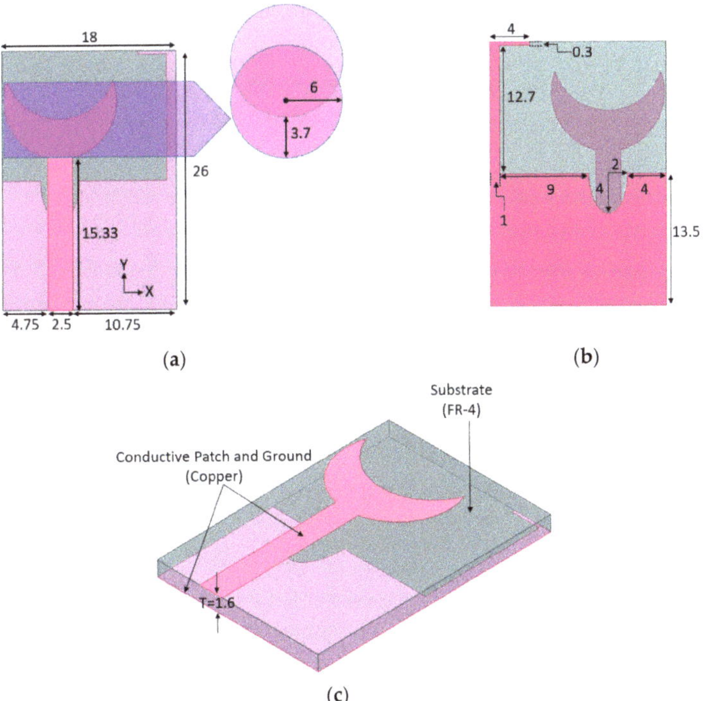

Figure 1. Single element antenna geometry (**a**) Top View (**b**) Bottom View (**c**) Perspective View (dimension in mm). Flame Retardant (FR).

The antenna is evaluated to check the effect on the reflection coefficient by varying the radius and position of the upper circle, as shown in Figure 2a,b. The variation of the upper circle radius (aa) (Figure 2a) affects the reflection coefficient levels majorly near the lower and upper part of the frequency band; however, the middle region is slightly affected. The value of the upper circular radius (aa) is chosen as 6 mm for wider bandwidth between the bands spanning from the semi-ellipse slot in terms of minor and major axis.

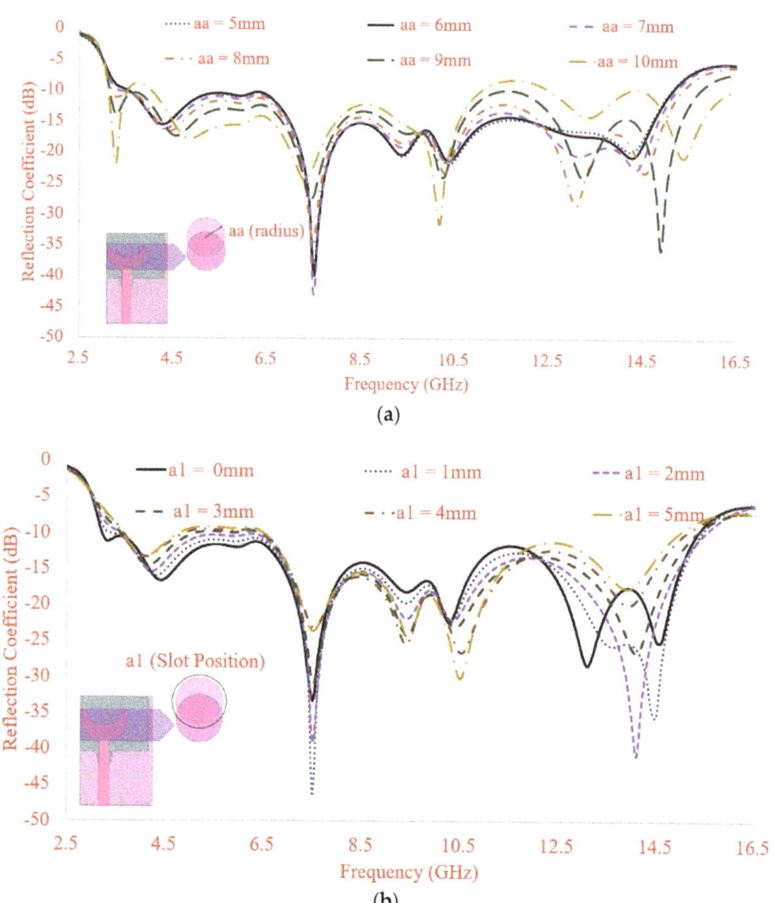

Figure 2. Parametric Variation of top circle (**a**) radius (aa) (**b**) position (a1).

The optimized value of the upper circle position helps in attaining the required reflection coefficient levels while achieving the wide impedance bandwidth. The variation in the position of the upper circle will ultimately increase or decrease the arc formed over the lower circle since the upper circle is subtracted from the lower one. The optimized position of the upper circle before subtracting it from the lower circle is selected as 0 mm.

By variation of the ground plane length as shown in Figure 3, it is observed that the impedance bandwidth significantly increases. The reflection coefficient of the resonating peaks is also affected. By increasing the length of the conductive region of the ground plane, the impedance bandwidth increases. The value of ground plane length is chosen as 13.5 mm for attaining the best performance.

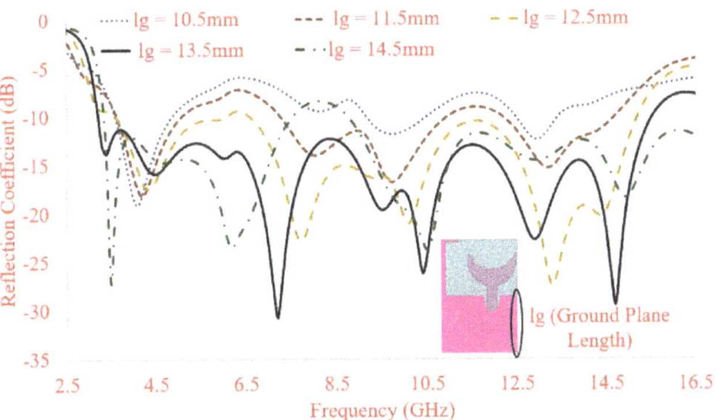

Figure 3. Parametric Variation of the ground plane (lg).a.

Figure 4 shows the top and back view of the basic antenna prototype fabricated on economical FR4 substrate (thickness, εr and tan (δ) of 1.524 mm, 4.5, and 0.002, respectively).

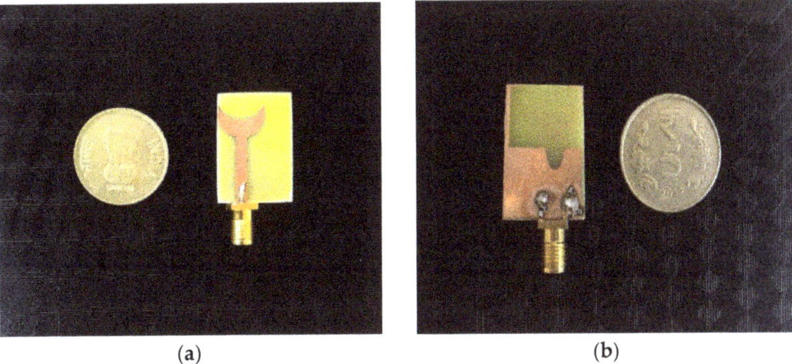

(a) **(b)**

Figure 4. Fabricated single element resonator (**a**) Front View (**b**) Back View.

The basic element is used for simulating a dual-element, quad-element, and eight-element UWB antenna. First, the single element is duplicated in the vertical and horizontal positions, and S parameters are analyzed. The element at port 2 is rotated in an anticlockwise direction concerning port 1 along the horizontal axis in case 1 (Figure 5a) while in case 2, the element is mirrored along the vertical axis (Figure 5b).

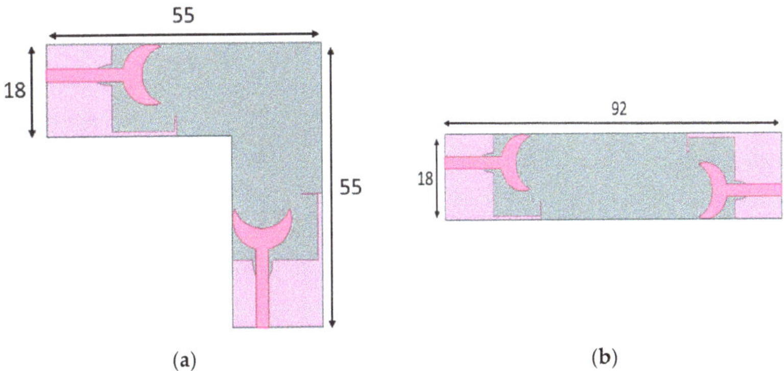

(a) **(b)**

Figure 5. Resonator geometry of 2-element MIMO (**a**) case 1 (**b**) case 2 (dimension in mm).

Four element MIMO design is realized by arranging the single element to form a plus-shaped structure. Orthogonal, as well as spatial diversity, along with separate ground is first attained as shown in Figure 6a. Since the applicability of the MIMO antenna with separate ground is very limited, the plus-shaped structure is modified to accomplish a connected ground 4-element MIMO. The antenna elements are symmetrically and rotationally placed in a manner so that it occupies low surface area while achieving pattern diversity. Concerning the element at port 1, the element at port 2 is rotated in an anticlockwise direction along the horizontal axis while the element at port 3 is mirrored along the vertical axis, and lastly, the element at port 4 is rotated clockwise along the horizontal axis. The connection between the ground planes is carefully chosen for isolation level below −15 dB, which is visible in Figure 6b.

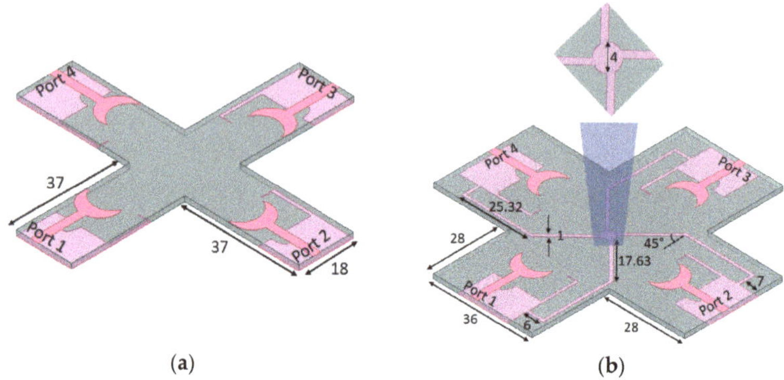

(a)

(b)

Figure 6. Resonator geometry of 4-element MIMO (**a**) Without common ground (**b**) with the common ground (dimension in mm).

The current geometry helps the antenna to be inserted in the systems where perfect square-shaped space is unavailable and can be accommodated easily between the circuit elements. The fabricated front and back view of the UWB 4-element connected ground MIMO antenna is illustrated in Figure 7a,b which is analyzed for MIMO diversity performance in the next section.

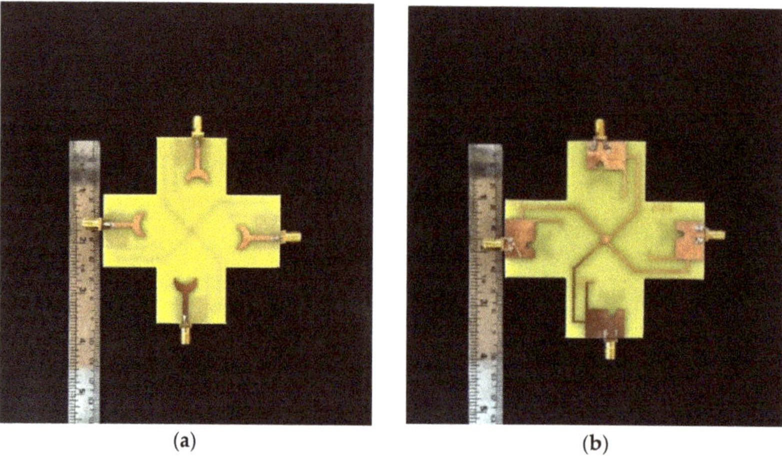

(a)

(b)

Figure 7. Fabricated 4-element Compact MIMO resonator (**a**) Front View (**b**) Back View (with the common ground).

Finally, an 8-element MIMO antenna is simulated as shown in Figure 8a,b where Figure 8a has a separate ground plane, and Figure 8b has connected ground. Additional four elements are added to the existing 4-element MIMO structure to achieve 8-element MIMO where two different cases are presented. Case 1 (Figure 8a) shows an 8-element structure without connected ground while in case 2 (Figure 8b), all the ground plane of all the elements is connected from the bottom side by carefully choosing the connecting lines.

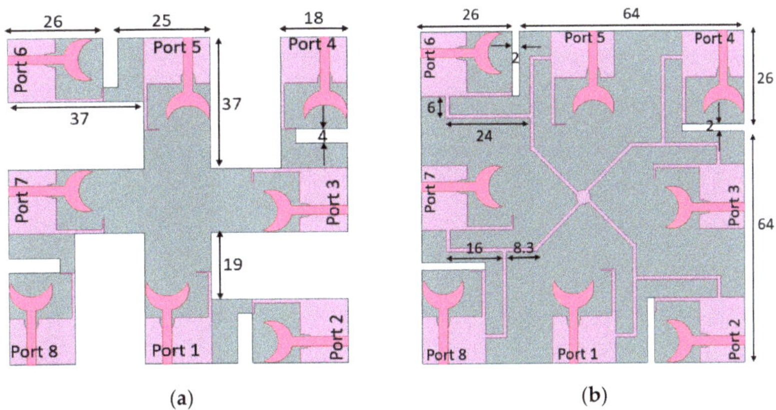

(a)   (b)

Figure 8. Resonator geometry of 8-element MIMO (**a**) Case 1, without common ground (**b**) Case 2, with common ground.

The analysis in terms of S parameters, current distribution, radiation patterns, gain, efficiency, and MIMO diversity parameters are discussed in the next section for the configurations shown here.

3. Results and Discussion

The reflection coefficient analysis of the basic element shows that the antenna covers an impedance bandwidth of (76.29%) 3.19–15.32 GHz where close correlation with simulated values is observed as illustrated in Figure 9.

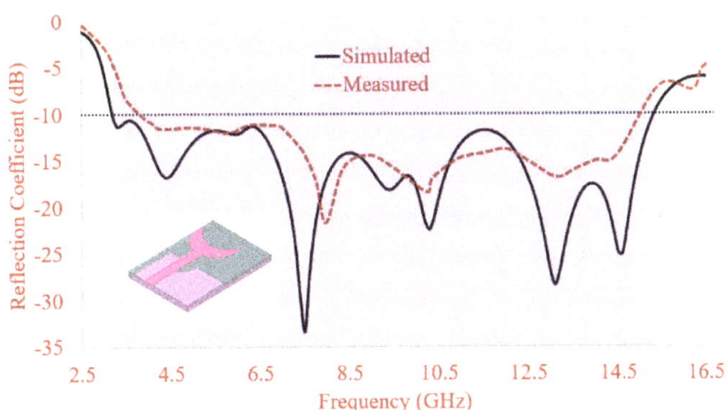

Figure 9. Simulated and Measured Reflection Coefficient (dB) of Single Element UWB Antenna.

The reflection coefficient of 2-element MIMO with separate ground is displayed in Figure 10 where apparent wide bandwidth and isolation >15 dB is observed for both the cases. The antenna resonates in the range of (134.70%) 3.01–15.43 GHz in case 1 while (134.38%) 3.01–15.34 GHz in case 2. The greater level of isolation is visible in the orthogonal arrangement of elements as compared to the elements arranged in a mirrored fashion along the vertical axis.

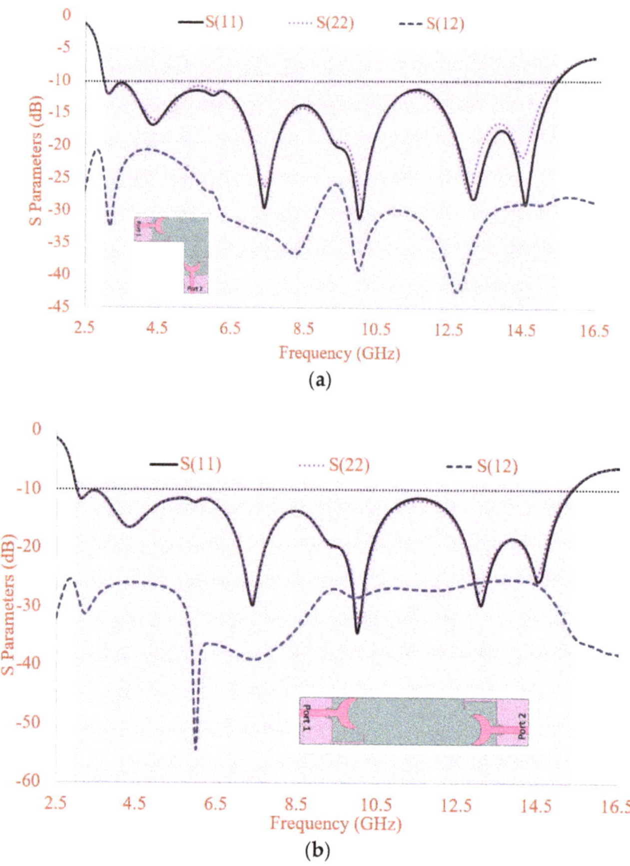

Figure 10. Simulated S Parameters of 2-Element Compact MIMO Antenna (dB) (**a**) Case 1 (**b**) Case 2.

The simulated reflection coefficient of four-element MIMO with the separate and connected ground plane is displayed in Figure 11 where the antenna achieves IBW of (135.25%) 2.98–15.43 GHz in antenna with separate ground and (136.92%) 2.84–15.33 GHz in antenna with the connected ground plane. Isolation levels are greater in the case of an antenna with separate ground; however, the careful selection of strip lines for connecting the ground planes has helped the antenna in achieving the isolation levels >15 dB due to minimum inter-element coupling. The measured results for connected ground configuration (Figure 12) indicate that over a wide frequency band spanning from (136.69%) 2.86–15.21 GHz, the results are well-matched. Slight variation in reflection coefficient and isolations levels are due to fabrication tolerances and connector losses.

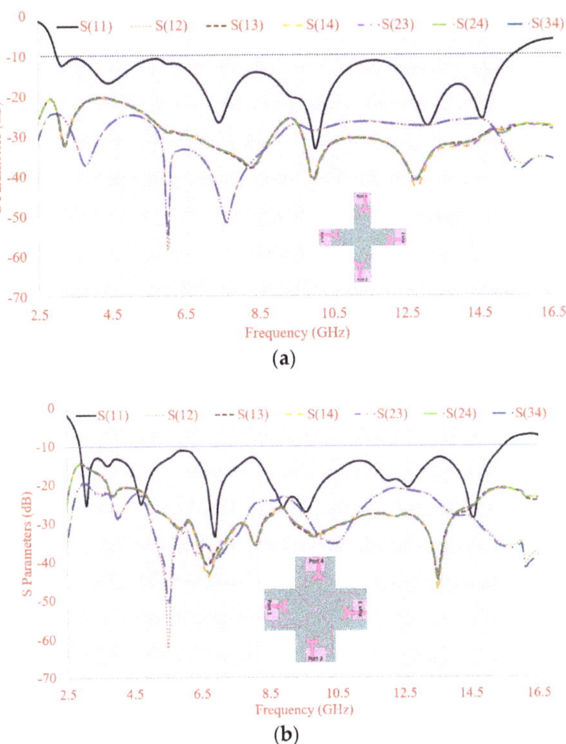

(a)

(b)

Figure 11. Simulated S Parameters of 4-Element Compact MIMO Antenna (dB) (**a**) without common ground (**b**) with common ground.

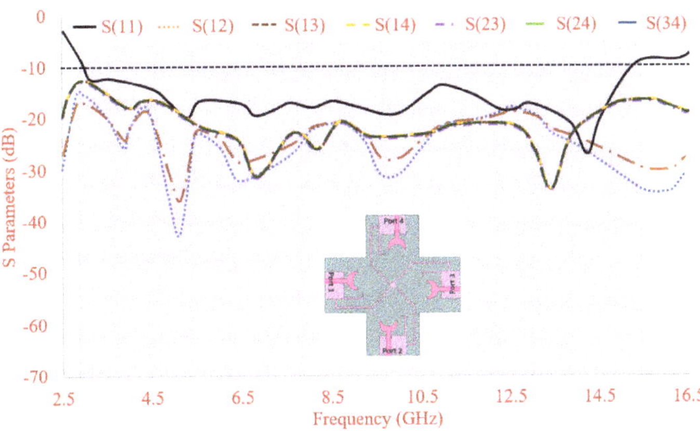

Figure 12. 4-Element Compact MIMO Antenna's experimental values of S Parameters with common ground.

Figure 12 shows the experimental values of 4-element-connected ground S parameters. The orthogonal placement of elements (1–2, 23, 34, and 41) leads to isolation >16 dB between the proposed bands. The isolation value is >12 dB is achieved between the elements placed in a mirrored fashion. It increases above 15 dB after 3.8 GHz. The orthogonal element placement ensures good isolation due to lower mutual coupling and the orthogonally polarized patterns. Higher isolation among elements placed in a mirrored position is due to the spatial separation. A decent agreement between simulated and experimental values is visible from Figure 12.

To comprehend the isolation mechanism, the current distribution is observed at 3.1 GHz, 4.70 GHz, 6.90 GHz, and 9.5 GHz in 4-element-connected ground structure as depicted in Figure 13a–d, respectively. While carrying out the analysis, Port 1 is excited while other ports are terminated with a 50 Ω load. The field intensity at four resonant peaks suggests that isolation is very good with a negligible amount of coupling occurring between the inter-elements.

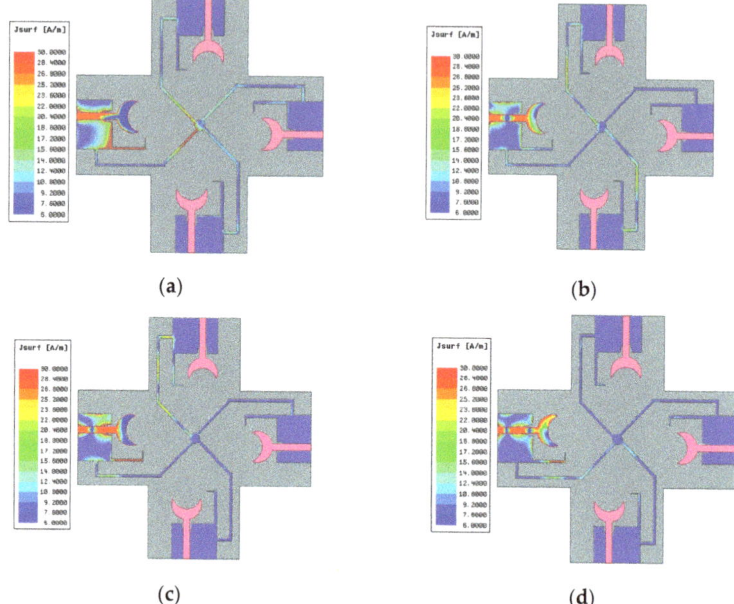

Figure 13. The current distribution of 4-element compact MIMO antenna with the common ground at (**a**) 3.10 GHz (**b**) 4.70 GHz (**c**) 6.90 GHz (**d**) 9.50 GHz.

The experimental two-dimensional (2D) co/cross-polarization radiation patterns of the UWB connected ground MIMO antenna are depicted in Figure 14 by providing excitation at port 1 and keeping other ports terminated with matched load (50 Ω). The EH plane patterns are plotted at resonating frequencies of 3.10 GHz, 4.70 GHz, 6.90 GHz, and 9.50 GHz. The lower cross-polarization at resonant peaks for both planes can be observed.

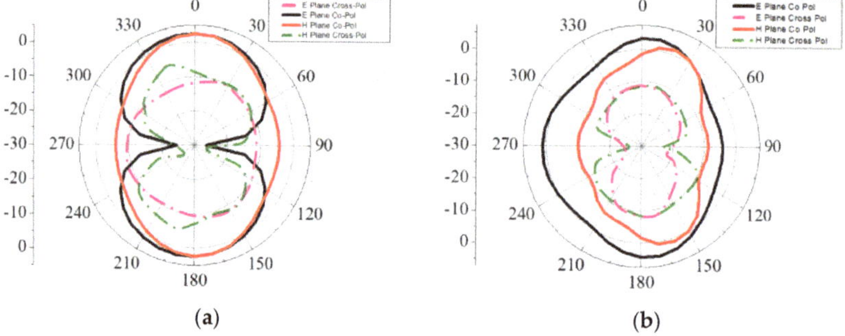

Figure 14. *Cont.*

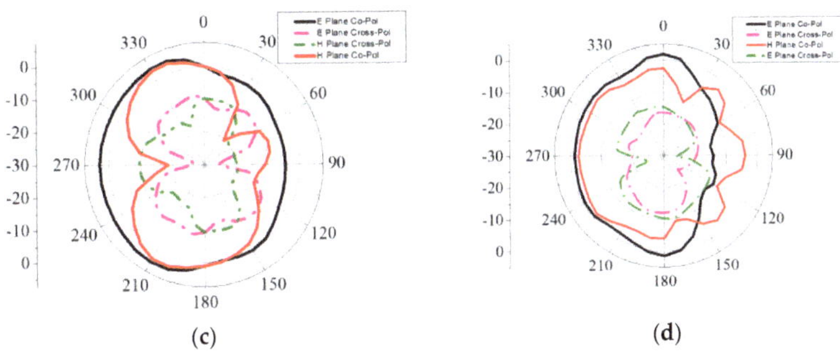

Figure 14. Measured Co/Cross Pol pattern of 4-element compact MIMO antenna with the common ground at (**a**) 3.10 GHz (**b**) 4.70 GHz (**c**) 6.90 GHz (**d**) 9.50 GHz.

Figure 15 depicts gain and efficiency plots by exciting port 1 and terminating other ports with impedance load (50 Ω). Average gain >4.5 dBi is observed for the bands of interest which closely matches with the measured gain values.

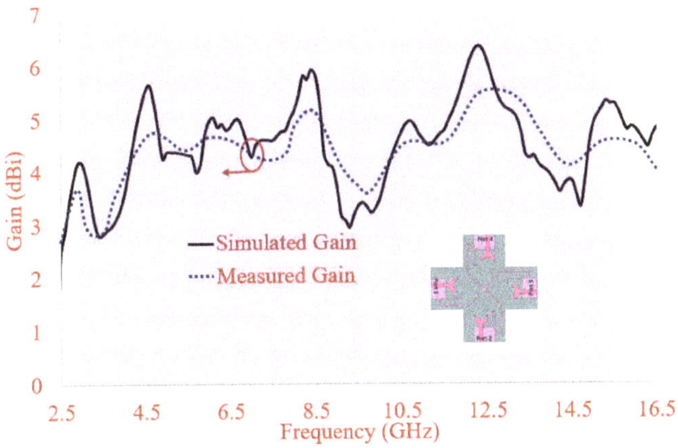

Figure 15. Gain and Efficiency of the 4-element MIMO antenna with common ground.

The S parameters of connected ground 8-element MIMO antenna are illustrated in Figure 16a–c. The simulated IBW of 8-port MIMO antenna is (137.47%)/2.84–15.3 as illustrated in (Figure 16a). Isolation of antenna concerning element 1 is depicted in Figure 16b where satisfactory isolation levels are observed having a value >15 dB, and the same is observed when isolation levels between other antenna elements are simulated (Figure 16c).

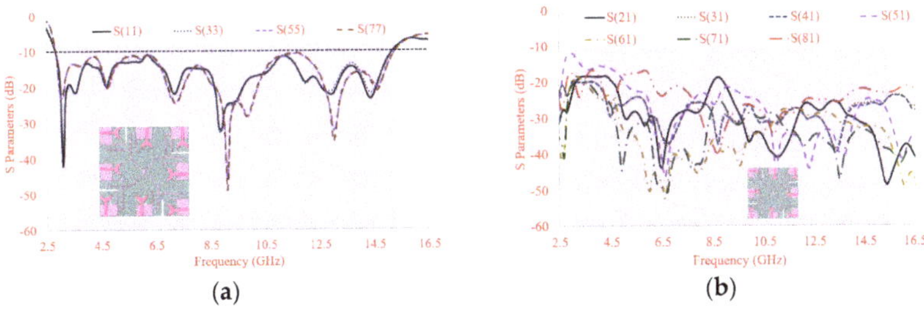

Figure 16. *Cont.*

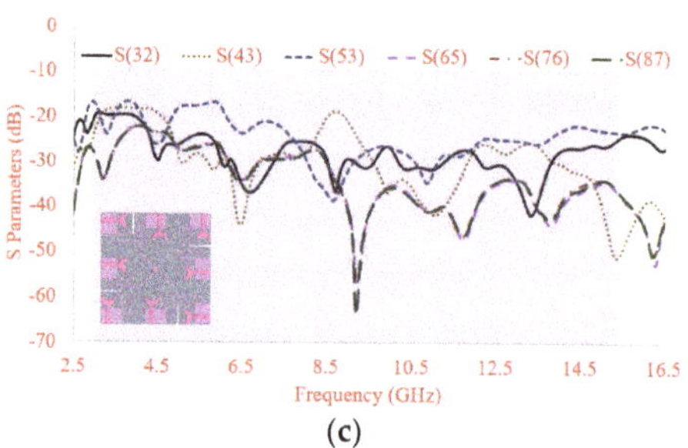

Figure 16. 8-Element Compact MIMO Antenna (dB) with common ground (**a**) Reflection coefficient (**b**) Isolation (antenna 1) (**c**) Isolation (Selected antennas).

A simulated gain and efficiency plot by exciting port 1 and terminating other ports with impedance load (50 Ω) for an 8-port MIMO antenna is illustrated in Figure 17. Average gain >5.6 dBi is observed for the bands of interest.

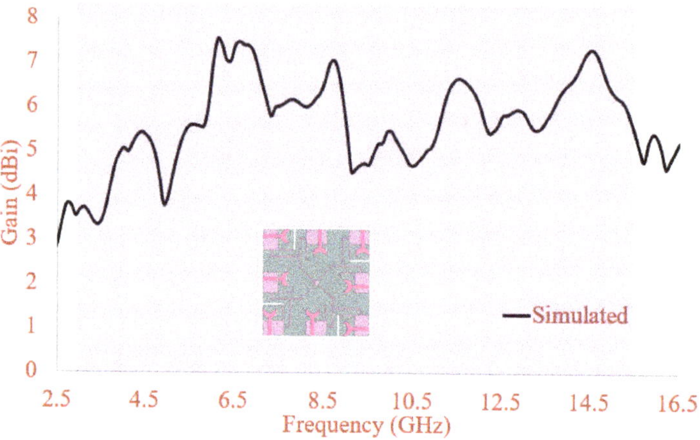

Figure 17. Gain and Efficiency of the 8-element MIMO antenna with common ground.

4. MIMO Diversity Analysis

Antenna characterization for diversity performance is carried out using the method suggested in [31–33] as the antenna is proposed for its use in MIMO applications.

4.1. Envelope Correlation Coefficient (ECC)

ECC is a very essential parameter for measuring the diversity performance of MIMO antenna. It indicates how radiations of each antenna element are independent of each other in a practical environment where the signals of transmitter and receiver are multipath fading signals.

The ECC of two antenna elements is estimated by using the below equation and graphically illustrated in Figure 18.

$$\rho_e = \frac{\left| \iint_{4\pi} \left[\vec{F_1}(\theta, \varnothing) * \vec{F_2}(\theta, \varnothing) \right] d\Omega \right|}{\iint_{4\pi} \left| \vec{F_1}(\theta, \varnothing) \right|^2 d\Omega \, \iint_{4\pi} \left| \vec{F_2}(\theta, \varnothing) \right|^2 d\Omega} \tag{1}$$

where $\vec{F_i}(\theta, \varnothing)$ is the three-dimensional field pattern of the antenna, when the *i*th port is excited. Ω is the solid angle measured over a sphere.

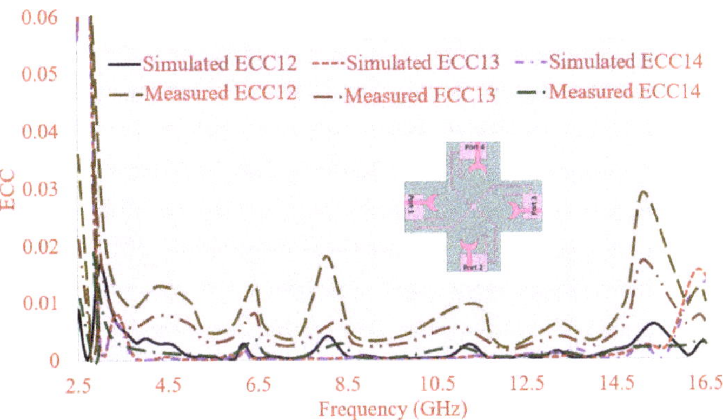

Figure 18. ECC of 4-element compact MIMO with common ground.

From Figure 18, it is noted that the ECC values (simulated and measured) of each antenna element (ECC 12, ECC 13, ECC 14) are very close to zero and less than 0.02. This validates that the MIMO antenna has uncorrelated far-field patterns and is suitable for UWB application.

4.2. Total Active Reflection Coefficient (TARC)

The TARC articulates the total incident power to the total outgoing power in presence of a multi-port antenna and is expressed in the below Equation (2) and shown in Figure 19.

$$TARC = \frac{\sqrt{\sum_{k=1}^{N}|b_k|^2}}{\sqrt{\sum_{k=1}^{N}|a_k|^2}} \qquad (2)$$

where in Equation (2), $|a|$ and $|b|$ are the excitation and scattering vector, respectively.

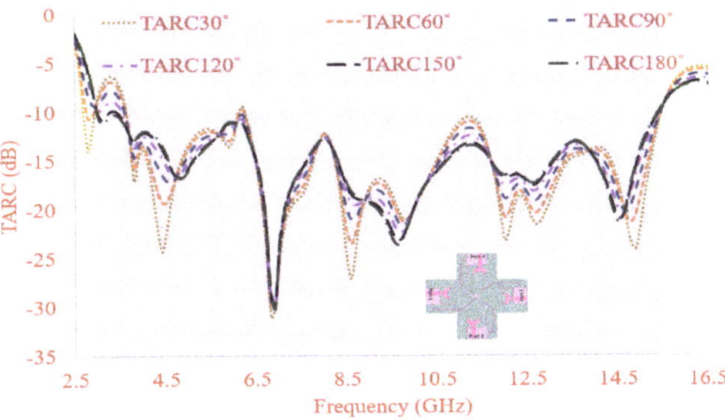

Figure 19. TARC of 4-element compact MIMO with common ground.

To verify the effect of TARC on impedance bandwidth, the proposed four-port MIMO antenna is integrated with an ideal phase shifter where a scan angle is varied from 30° to 180°. From Figure 18, it is visualized that for all scanning angles the TARC value lies below −10 dB confirming that all the delivered power is accepted by another antenna element without affecting the impedance bandwidth of the proposed four-port MIMO antenna. Therefore, this also validated that the proposed four-port MIMO antenna is also an attractive applicant for integration with a phase shifter.

4.3. Mean Effective Gain (MEG)

The MEG is another essential diversity performance parameter for the characterization of MIMO antenna in wireless channels. MEG determines the ability of the antenna element to accept an electromagnetic signal in the presence of affluent multipath fading signals.

Therefore, Figure 20 illustrates the ratio of the mean received power to mean incident power of antenna elements. Each MEG ratio closer to 0 dB throughout the operating bands confirms that each antenna element is a suitable candidate for UWB MIMO applications.

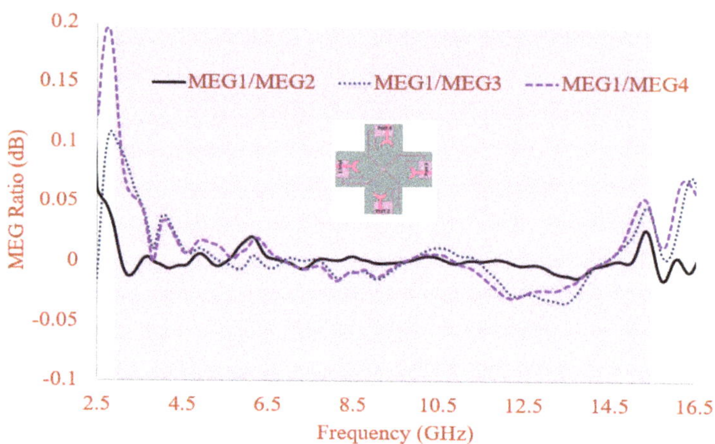

Figure 20. MEG Ratio of 4-element compact MIMO with common ground.

4.4. Channel Capacity Loss (CCL)

To achieve high data transmission, the minimum acceptable level of CCL is 0.5 bits/s/Hz. The CCL is calculated using Equation (3) and shown in Figure 21. The CCL values are as low as 0.4 bits/s/Hz which is well within the requirement set by the industry and matches well with the experimental values.

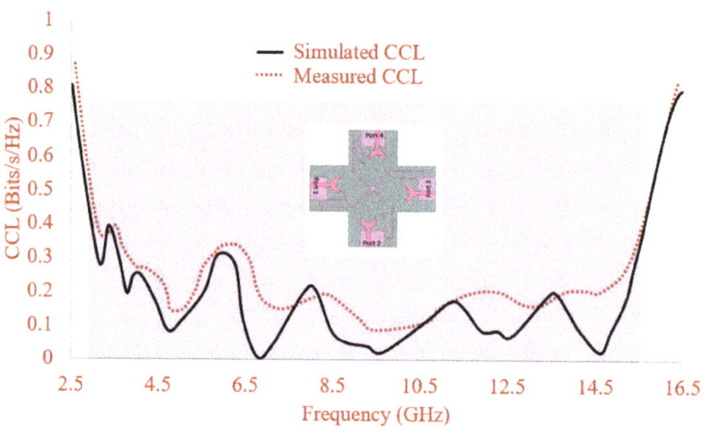

Figure 21. CCL of 4-element compact MIMO with common ground.

4.5. ECC of 8-Element-Connected Ground MIMO Antenna

From Figure 22, it is noted that the simulated ECC values of each antenna element (ECC 12, ECC 13, ECC 14, ECC 15, ECC 16, ECC 17, and ECC 18) are very close to zero and less than 0.02. This validates that the MIMO antenna has uncorrelated far-field patterns.

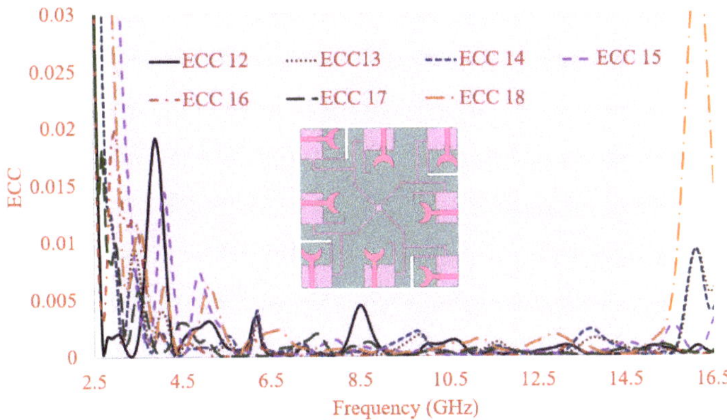

Figure 22. ECC of 8-element compact MIMO with common ground.

5. Time Domain Performance Analysis of the Proposed Antenna

In order to illustrate the time domain performance of the antenna, the time domain characteristics including forward transmission (S_{21}) coefficient and group delay are investigated.

Figure 23 shows the experimental set-up that uses twin antennas, where one is acting as a transmitter (Tx) and another as receiver (Rx). These antennas are placed face-to-face at a distance of 50 cm which is five times the wavelength of lower operating frequency (around 3.00 GHz) in order to create far field atmosphere. Twin antennas are excited by using 5th order Gaussian pulse by using following Equation (3):

$$x(t) = A\left(-\frac{15t}{\sqrt{2\pi}\sigma^7} + \frac{10t^3}{\sqrt{2\pi}\sigma^9} \frac{t^5}{\sqrt{2\pi}\sigma^{11}}\right) x \exp\left(-\frac{t^2}{2\sigma^2}\right) \tag{3}$$

where in (3), 't' is time, 'A' is amplitude, 'σ' is spread of Gaussian pulse.

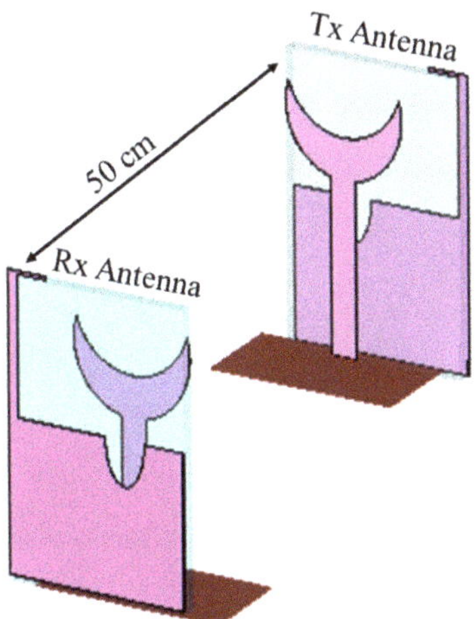

Figure 23. Test set-up for time domain analysis of the proposed antenna in CST MWS.

In order to check the time domain response of the proposed antenna, the transmission coefficient (S_{21}) of experimental set-up was simulated and shown in Figure 24. It is inspected that, proposed antenna exhibits the magnitude of the transmission coefficient (S_{21}) below −25 dB over the entire operating bands.

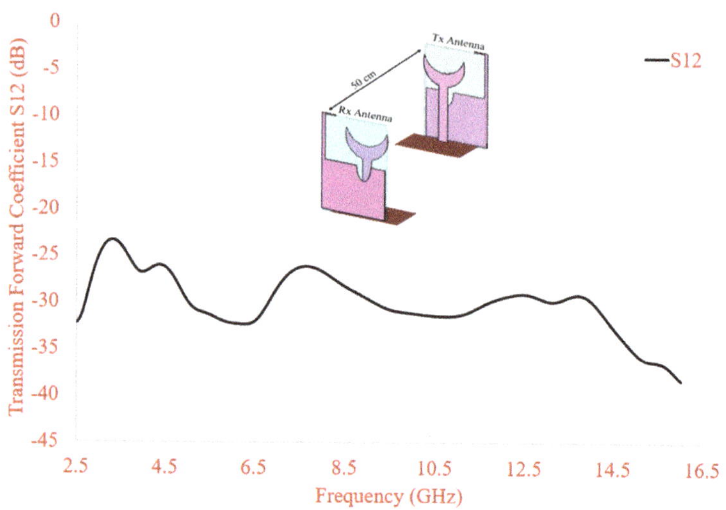

Figure 24. Transmission coefficient of the proposed antenna.

Figure 25 shows the simulated group delay response of experimental set-up. It is observed that the group delay remains stable around 1.5 ns in operating bands of the proposed antenna.

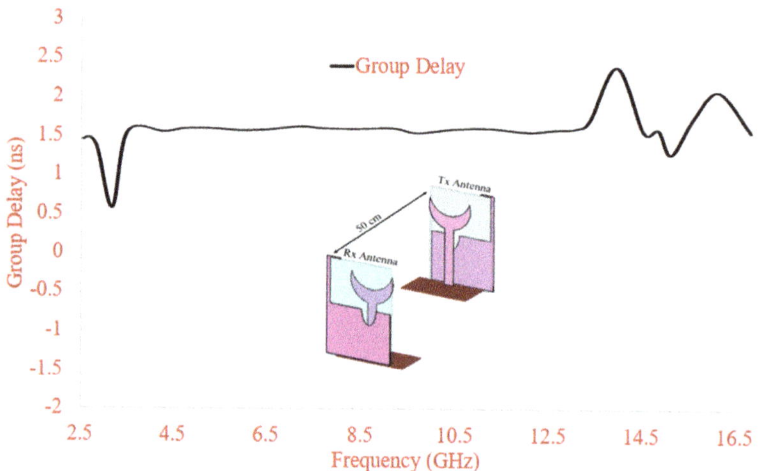

Figure 25. Group delay of the presented antenna.

Through the investigation of the time domain behavior of the proposed antenna, it can be concluded that the proposed antenna has linear transmission characteristics with lower transmission coefficient and stable group delay across the operating band's interest.

This confirms the applicability of the proposed antenna for UWB operations employed in wireless devices.

Table 1 shows the comparison of the proposed 4/8-element UWB antenna with other UWB MIMO antennas. The proposed 4/8-element antenna exhibits low profile as compared to most of the 8-port MIMO antennas, high IBW, isolation >15 dB without using any complex isolation or decoupling techniques, high gain as compared to all the existing state of Arts, and permissible MIMO diversity performance that makes the antenna commercially viable for portable UWB applications.

Table 1. Comparison of proposed 4/8 elements UWB MIMO antenna with other UWB MIMO antennas.

Ref. No	No of Elements	Size (mm^2) (At Lowest Freq)	Operating Band (GHz)	Isolation (dB)	Isolating Element	Peak Gain (dBi)	ECC
[12]	8	0.41 λ × 0.41 λ	3.1–11	>20	Not Used	4	<0.01
[13]	4	0.41 λ × 0.44 λ	3.1–10.6	>20	Microstrip Multimode Resonator	4	<0.2
[15]	4	0.77 λ × 0.77 λ	3.1–17.3	>13.5	Neutralization Ring	5.6	<0.1
[16]	4	0.67 λ × 0.67 λ	2.8–13.3	>18	Rectangular Strip	6	<0.06
[17]	4	0.15 λ × 0.3 λ	0.75–7.65	>12	Not used	3.2	<0.24
[18]	4	0.45 λ × 0.26 λ	3.2–12	>22	Not used	4	<0.5
[20]	4	0.56 λ × 0.39 λ	(3.52–10.08)	>22	Neutralization Line	2.91	<0.04
[21]	4	0.44 λ × 0.44 λ	3.8–15	>15	Ground Stubs	3	<0.07
[26]	8	1.65 λ × 0.82 λ	4.95−5.05	>10.5	Spatial Diversity	0.8	–
[27]	8	1.24 λ × 0.62 λ	3.4−3.6	>10	CL-FSS, circular arcs, and four parallel strips	–	–
[28]	8	0.6 λ × 0.93 λ	3−10.6	>15	square loop radiating strip with orthogonal polarisation	–	–
[29]	8	0.56 λ × 1.13 λ	2.5−2.6	>15	Eight grounded Slits	0.7	<0.2
[30]	8	0.75 λ × 1.5 λ	3.3−6	>11	Not used	–	<0.1
[31]	8	0.85 λ × 0.85 λ	3−10.6	>15	Rectangular stub	4.8	<0.2
Proposed	8	0.61 λ × 0.61 λ	2.84–11 (effective BW)	>15	Not Used	7.2	<0.02

6. Conclusions

A four-element connected ground UWB-MIMO antenna with modified substrate geometry, and the defected ground is presented in this article. The low profile of the antenna is achieved using substrate shape which is different than the conventional square shape. The key parameters of the 4-element antenna are analyzed using simulation and experiment in terms of |S11|, radiation patterns, isolation, gain, and MIMO diversity parameters. A decent agreement is observed between the results. Connected ground helps the MIMO antenna for its use in commercial applications moreover defected structure with inverted L-shaped strip, and slotted semi ellipse at the ground plane helped in accomplishing the wideband performance. Isolation between elements >16 dB is achieved by orthogonally placing the antenna elements. The proposed low profile 4-elements UWB-MIMO antenna structure achieves wide IBW, stable gain, omnidirectional patterns, higher isolation, and decent diversity properties. The extended connected ground 8-element

MIMO antenna also illustrated wide IBW, satisfactory isolation levels, and ECC that makes it a good candidate for portable UWB-MIMO systems.

Author Contributions: Conceptualization, T.A., A.D. and I.E.; methodology, T.A., N.A., J.K.; software, T.A., A.D.; validation, i.e., C.Z., and N.A.; formal analysis, A.D.; investigation, T.A., i.e., J.K.; resources, J.R., C.Z.; writing—original draft preparation, T.A., A.D.; writing—review and editing, i.e., R.A.-A., N.A.; visualization, J.K., C.Z.; supervision, R.A.-A.; project administration, J.R.; funding acquisition, I.E. All authors have read and agreed to the published version of the manuscript.

Funding: This work is funded by FCT/MCTES through national funds and when applicable co-funded EU funds under the project UIDB/50008/2020-UIDP/50008/2020.

Data Availability Statement: All data are included within manuscript.

Acknowledgments: This work is funded by FCT/MCTES through national funds and when applicable co-funded EU funds under the project UIDB/50008/2020-UIDP/50008/2020. Authors Issa Elfergani and Jonathan Rodriguez would like to acknowledge the POSITION-II project (ECSEL joint Undertaking under grant number Ecsel-7831132-Position-II-2017-IA).

Conflicts of Interest: The authors declare no conflict of interest.

References

1. Federal Communications Commission. *Federal Communications Commission Revision of Part 15 of the Commision's Rules Regarding Ultra-Wideband Transmission System from 3.1 to 10.6 GHz*; ET-Docket: Washington, DC, USA, 2002.
2. Kaiser, T.; Zheng, F.; Dimitrov, E. An overview of ultrawide- band systems with MIMO. *Proc. IEEE* **2009**, *97*, 285–312. [CrossRef]
3. Sharawi, M.S. Current misuses and future prospects for printed multiple input, multiple-output antenna systems [wireless corner]. *IEEE Antennas Propag. Mag.* **2017**, *59*, 162–170. [CrossRef]
4. Watcharaphon, N.; Ruengwaree, A. Four-Port Rectangular Monopole Antenna for UWB-MIMO Applications. *Prog. Electromagn. Res.* **2020**, *87*, 19–38.
5. Kulkarni, J.; Desai, A.; Sim, C.-Y.-D. Wideband four-port MIMO antenna array with high isolation for future wireless systems. *Int. J. Electron. Commun.* **2021**, *128*, 1–15. [CrossRef]
6. Wu, W.; Yuan, B.; Wu, A. A quad-element UWB-MIMO antenna with band-notch and reduced mutual coupling based on EBG structures. *Int. J. Antennas Propag.* **2018**. [CrossRef]
7. Amin, F.; Saleem, R.; Shabbir, T.; Bilal, M.; Shafique, M.F. A compact quad-element UWB-MIMO antenna system with parasitic decoupling mechanism. *Appl. Sci.* **2019**, *9*, 2371. [CrossRef]
8. Alam, T.; Thummaluru, S.R.; Chaudhary, R.K. Integration of MIMO and Cognitive Radio for Sub-6 GHz 5G Applications. *IEEE Antennas Wirel. Propag. Lett.* **2019**, *18*, 2021–2025. [CrossRef]
9. Anitha, R.; Vinesh, P.V.; Prakash, K.C.; Mohanan, P.; Vasudevan, K. A compact quad element slotted ground wideband antenna for MIMO applications. *IEEE Trans. Antennas Propag.* **2016**, *64*, 4550–4553. [CrossRef]
10. Mathur, R.; Dwari, S. Compact 4-Port MIMO/diversity antenna with low correlation for UWB application. *Frequenz* **2018**, *72*, 429–435. [CrossRef]
11. Tripathi, S.; Mohan, A.; Yadav, S. A compact Koch fractal UWB MIMO antenna with WLAN band-rejection. *IEEE Antennas Wirel. Propag. Lett.* **2015**, *14*, 1565–1568. [CrossRef]
12. Sipal, D.; Abegaonkar, M.; Koul, S. Easily extendable compact planar UWB MIMO antenna array. *IEEE Antennas Wirel. Propag. Lett.* **2017**, *16*, 2328–2331. [CrossRef]
13. Ali, W.; Ibrahim, A.A. A compact double-sided MIMO antenna with an improved isolation for UWB applications. *Int. J. Electron. Commun.* **2017**, *82*, 7–13. [CrossRef]
14. Rekha, V.; Pokkunuri, P.; Boddapati, T.; Madhav, P.; Yalavarthi, U. Dual Band Notched Orthogonal 4-Element MIMO Antenna with Isolation for UWB Applications. *IEEE Access* **2020**, *8*, 145871–145880. [CrossRef]
15. Kayabasi, A.; Toktas, A.; Yigit, E.; Sabanci, K. Triangular quad-port multi-polarized UWB MIMO antenna with enhanced isolation using neutralization ring. *Int. J. Electron. Commun.* **2018**, *85*, 47–53. [CrossRef]
16. Kumar, S.; Gwan, H.; Dong, H.; Wahab, M.; Hyun, C.; Kang, K. Multiple-input-multiple-output/diversity antenna with dual band-notched characteristics for ultra-wideband applications. *Microw. Opt. Technol. Lett.* **2020**, *62*, 336–345. [CrossRef]
17. Hussain, R.; Sharawi, M.S.; Shamim, A. An Integrated Four-Element Slot-Based MIMO and a UWB Sensing Antenna System for CR Platforms. *IEEE Trans. Antennas Propag.* **2018**, *66*, 978–983. [CrossRef]
18. Srivastava, G.; Mohan, A. Compact MIMO slot antenna for UWB applications. *IEEE Antennas Wirel. Propag. Lett.* **2015**, *15*, 1057–1060. [CrossRef]
19. Sarkar, D.; Srivastava, K.V. A compact four-element MIMO/diversity antenna with enhanced band width. *IEEE Antennas Wirel. Propag. Lett.* **2017**, *16*, 2469–2472. [CrossRef]
20. Tiwari, R.; Singh, P.; Kanaujia, B.; Srivastava, K. Neutralization technique based two and four port high isolation MIMO antennas for UWB communication. *Int. J. Electron. Commun.* **2019**, *110*, 152828. [CrossRef]

21. Zamir, W.; Kumar, D. A compact 4× 4 MIMO antenna for UWB applications. *Microw. Opt. Technol. Lett* **2016**, *58*, 1433–1436.
22. Li, W.; Yongqiang, H.; Grubb, P.; Shi, X.; Ray, C. Compact inkjet-printed flexible MIMO antenna for UWB applications. *IEEE Access* **2018**, *6*, 50290–50298. [CrossRef]
23. Desai, A.; Upadhyaya, T.; Palandoken, M.; Gocen, C. Dual band transparent antenna for wireless MIMO system applications. *Microw. Opt. Technol. Lett.* **2019**, *61*, 1845–1856. [CrossRef]
24. Desai, A.; Cong, B.; Patel, J.; Upadhyaya, T.; Byun, G.; Nguyen, T. Compact Wideband Four Element Optically Transparent MIMO Antenna for mm-Wave 5G Applications. *IEEE Access* **2020**, *8*, 194206–194217. [CrossRef]
25. Sharawi, M.S. A 5-ghz 4/8-element MIMO antenna system for IEEE 802.11 ac devices. *Microw. Opt. Technol. Lett.* **2013**, *55*, 1589–1594. [CrossRef]
26. Al-Hadi, A.A.; Ilvonen, J.; Valkonen, R.; Viikari, V. Eight-element antenna array for diversity and MIMO mobile terminal in LTE 3500 MHz band. *Microw. Opt. Technol. Lett.* **2014**, *56*, 1323–1327. [CrossRef]
27. Saleem, R.; Bilal, M.; Bajwa, K.; Shafique, M. Eight-element uwb-mimo array with three distinct isolation mechanisms. *Electron. Lett.* **2015**, *51*, 311–313. [CrossRef]
28. Li, M.; Xu, Z.; Ban, Y.; Sim, C.; Yu, Z. Eight-port orthogonally dual-polarised MIMO antennas using loop structures for 5g smartphone. *IET Microw. Antennas Propag.* **2017**, *11*, 1810–1816. [CrossRef]
29. Mathur, R.; Dwari, S. 8-port multibeam planar uwb-mimo antenna with pattern and polarisation diversity. *IET Microw. Antennas Propag* **2019**, *13*, 2297–2302. [CrossRef]
30. Zhang, X.; Li, Y.; Wang, W.; Shen, W. Ultra-wideband 8-port MIMO antenna array for 5g metal-frame smartphones. *IEEE Access* **2019**, *7*, 72273–72282. [CrossRef]
31. Sharawi, M.S. Printed multi-band MIMO antenna systems and their performance metrics [wireless corner]. *IEEE Antennas Propag. Mag.* **2013**, *55*, 218–232. [CrossRef]
32. Najam, A.I.; Duroc, Y.; Tedjini, S. *Multiple-Input Multiple-Output Antennas for Ultra-Wideband Communications*; IntechOpen: Rijeka, Croatia, 2012; Volume 10, pp. 209–236.
33. Elfergani, I.T.E.; Hussaini, A.S.; Rodriguez, J.; Abd-Alhameed, R. *Antenna Fundamentals for Legacy Mobile Applications and Beyond*; Springer: Cham, Switzerland, 2017; pp. 1–659.

 *electronics*

Article

Miniaturized Broadband-Multiband Planar Monopole Antenna in Autonomous Vehicles Communication System Device

Ming-An Chung * and Chih-Wei Yang

Department of Electronic Engineering, National Taipei University of Technology, Taipei City 10608, Taiwan;
t109368113@ntut.org.tw
* Correspondence: mingannchung@ntut.edu.tw; Tel.: +886-2-2771-2171 (ext. 2212)

Abstract: The article mainly presents that a simple antenna structure with only two branches can provide the characteristics of dual-band and wide bandwidths. The recommended antenna design is composed of a clockwise spiral shape, and the design has a gradual impedance change. Thus, this antenna is ideal for applications also recommended in these wireless standards, including 5G, B5G, 4G, V2X, ISM band of WLAN, Bluetooth, WiFi 6 band, WiMAX, and Sirius/XM Radio for in-vehicle infotainment systems. The proposed antenna with a dimension of 10×5 mm is simple and easy to make and has a lot of copy production. The operating frequency is covered with a dual-band from 2000 to 2742 MHz and from 4062 to beyond 8000 MHz and, it is also demonstrated that the measured performance results of return loss, radiation, and gain are in good agreement with simulations. The radiation efficiency can reach 91% and 93% at the lower and higher bands. Moreover, the antenna gain can achieve 2.7 and 6.75 dBi at the lower and higher bands, respectively. This antenna design has a low profile, low cost, and small size features that may be implemented in autonomous vehicles and mobile IoT communication system devices.

Keywords: dual-band; mobile IoT; autonomous vehicles; B5G; 5G; V2X; DSRC; WiFi 6 band; Sirius/XM Radio; ISM band

Citation: Chung, M.-A.; Yang, C.-W. Miniaturized Broadband-Multiband Planar Monopole Antenna in Autonomous Vehicles Communication System Device. *Electronics* **2021**, *10*, 2715. https://doi.org/10.3390/electronics10212715

Academic Editors: Issa Tamer Elfergani, Raed A. Abd-Alhameed and Abubakar Sadiq Hussaini

Received: 13 October 2021
Accepted: 5 November 2021
Published: 8 November 2021

Publisher's Note: MDPI stays neutral with regard to jurisdictional claims in published maps and institutional affiliations.

1. Introduction

The vigorous development of IoT technology has laid a solid foundation for continued growth in the 5G era [1,2]. Therefore, networking, artificial intelligence, virtual reality (VR), and augmented reality (AR) technology can be used to provide relevant local 5G-centered services in health care, education, and other fields [3,4]. The rise of 5G communications has initiated various IoT business model development. The driving factors that trigger IoT applications include the low cost of storing and computing data on cloud platforms. In addition, it also includes emerging edge computing trends, the decline in data, sensors, equipment costs, and the availability of mobile application development platforms [5]. Therefore, emerging technologies integrate intelligent roads, intelligent vehicles, and artificial intelligence into our lives [6]. Therefore, 5G technology still has a lot to be discussed and technical improvement, combined with other wireless communication technologies, and continues to improve wireless communication development in Beyond 5G (B5G). Large-scale low-latency internet of things and 5G private network services focus on the next generation of B5G/6G development. Therefore, the spectrum planning and telecommunication service mode of B5G satellite communications will be essential in the future. The use of low-orbit satellites is expected to supplement areas that ground base stations cannot cover and through 5G/B5G/emerging wireless communication, vertical application demonstrations, IoT devices, and scenarios are introduced to provide innovative applications [7,8].

Furthermore, low-orbit satellite communication is also a key technology for future 6G commercial transformation. The development of 5G technology to 6G and satellite communications also includes the evolution of existing technologies and the development

151

of emerging technologies for terrestrial communications, which will generate demand for spectrum planning [7,9]. In addition, the rapid progress of telecommunications infrastructure will affect the future development of artificial intelligence and self-driving vehicles.

The current internet of things devices must have a good match with wireless communication. However, the antenna design requires a design with multiple frequency bands and a large bandwidth is the same basic design principle. The main reason is that various wireless communication design standards must be accommodated in one device [10–12].

In recent literature discussion, antenna design is mostly compact multi-band antenna design as the main discussion topic. The superior design of coplanar waveguide (CPW) and slot line antenna (SL) can be used to achieve antenna design goals, including easy manufacturing and high compatibility with microwave circuits. In addition, using the CPW architecture design, CPW line segments with different widths and gap widths are used to achieve the ideal resonance frequency and massive bandwidth [13–15].

Multi-band antennas have integrated the applications of multiple wireless communication standards. It can be seen that the antenna design of IoT mobile devices with multi-band operation capabilities will be in great demand in the future market. Therefore, many design and development methods have been extended to multi-band antennas and integrated into various wireless communication product applications. However, the antenna design can be implemented in multiple frequency bands with antenna size minimization. Different generating frequency bands have been proposed in a single antenna, such as double L-slit, inductive slot [16–25].

The design of IoT electronic products requires miniaturization and compact circuit design. Therefore, the antenna design space must be sacrificed. However, the antenna's effectiveness still needs to reach a certain level—no matter what kind of wireless communication product, finding any possible design practice is an important issue.

This research article mainly presented a miniaturized broadband-multiband planar monopole antenna with a clockwise spiral shape of two branches for Beyond-5G, 5G, 4G, V2X, DSRC, WiFi 6 band, WLAN, and WiMAX application in autonomous vehicles and mobile IoT communication system device.

2. Recommended Monopole Antenna Construction

The recommended antenna structure is disclosed and presented in Figure 1—the proposed antenna design with the best characteristic, including small size, broadband bandwidth, and low-profile structure. The proposed antenna employed a clockwise spiral shape of two branches to achieve dual-band performance covering 2000 to 2742 MHz and 4062 to over 8000 MHz, as shown in Figure 1a,e. The proposed antenna is validated by a mini coaxial cable with the characteristic impedance of 50 ohms and connects to an SMA connector, as shown in Figure 1a. This design is suitable for eight novel wireless communication systems, including Beyond-5G for LEO Device to Device (D2D) application, 5G, 4G, V2X, DSRC, WiFi 6 band, WLAN, and WiMAX systems. The antenna material is simulated and fabricated using the most commonly used FR4 substrate in the industry, as shown in Figure 1b. The antenna area is placed on the edge side in a device design, and the related wireless and baseband function components for the vehicle applications will be designed in the ground plane to combine the antenna design. The design parameters of the FR4 substrate are with the dielectric constant (εr) of 4.4, the dielectric loss tangent of 0.00245, and the height of 0.8mm.

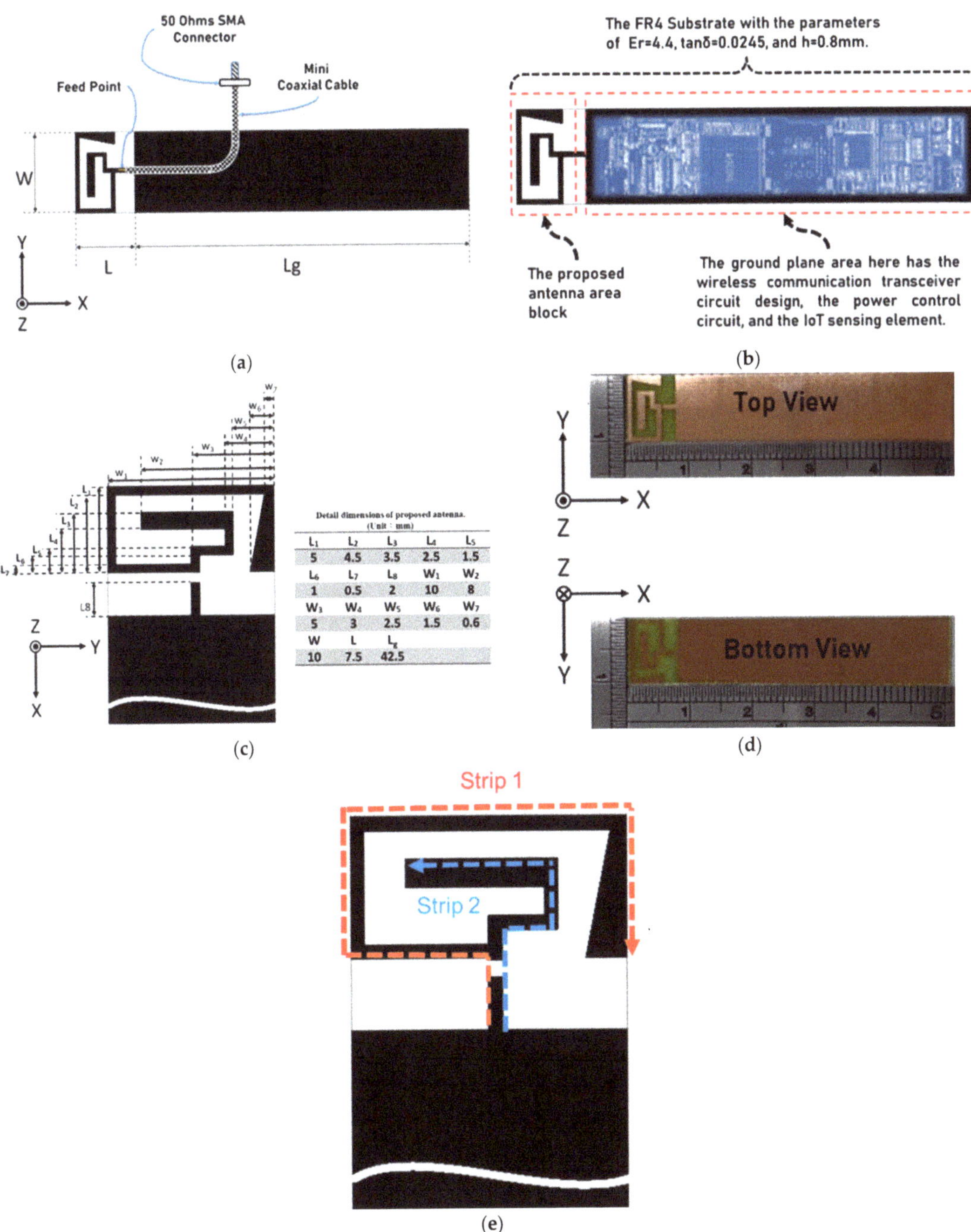

Detail dimensions of proposed antenna. (Unit : mm)

L_1	L_2	L_3	L_4	L_5
5	4.5	3.5	2.5	1.5
L_6	L_7	L_8	W_1	W_2
1	0.5	2	10	8
W_3	W_4	W_5	W_6	W_7
5	3	2.5	1.5	0.6
W	L	L_g		
10	7.5	42.5		

Figure 1. Disclosure of recommended antenna structure: (**a**) view of the antenna structure and overall design of the measurement layout, (**b**) schematic diagram of the layout applied to IoT devices; (**c**) detailed dimensions and layout of the antenna; (**d**) antenna entity diagram; (**e**) the clockwise spiral shape of two branches (Strip 1 and Strip 2).

The main antenna area with the dimension of 10×7.5 mm, and the ground for system placement includes an integrated circuit with power, baseband, and wireless transceiver function, which is located the dimension of 10×42.5 mm, as the detailed information displayed in Figure 1a–c. The actual antenna photo produced is shown in Figure 1d. The antenna design only needs the top surface to complete the design. Therefore, the tolerance sensitivity of this design is low, and it is easy to adjust the accurate operating frequency design. In addition, the clockwise spiral shape of two branches has defined Strip 1 and Strip 2 to explain the physical phenomena of surface current distribution analysis and resonance frequency, as shown in Figure 1e.

The surface current distribution is analyzed in Figure 2. The lower band is displayed at 2350 and 2450 MHz in Figure 2a,b. The upper band at 5080, 5500, 5890, 6500, and 7150 MHz are in Figure 2c–g. There is a strong current on the branch of Strip 1 at the operation frequency of 2350 and 2450 MHz. Therefore, it can be clearly understood that the operating frequency in the low-frequency band is a quarter wavelength contributed by Strip 1. Strip 2 mainly excites resonance at operating frequencies of 5080 and 5500 MHz, as shown in Figure 2c,d of the current distribution diagram. Finally, the operation frequency over 5890 MHz is affected by the parallel coupling path of Strip 1, and Strip 2 produces a high-frequency broadband effect. The effect of the parallel coupling path of Strip 1 and Strip 2 is demonstrated in Figure 2e–g.

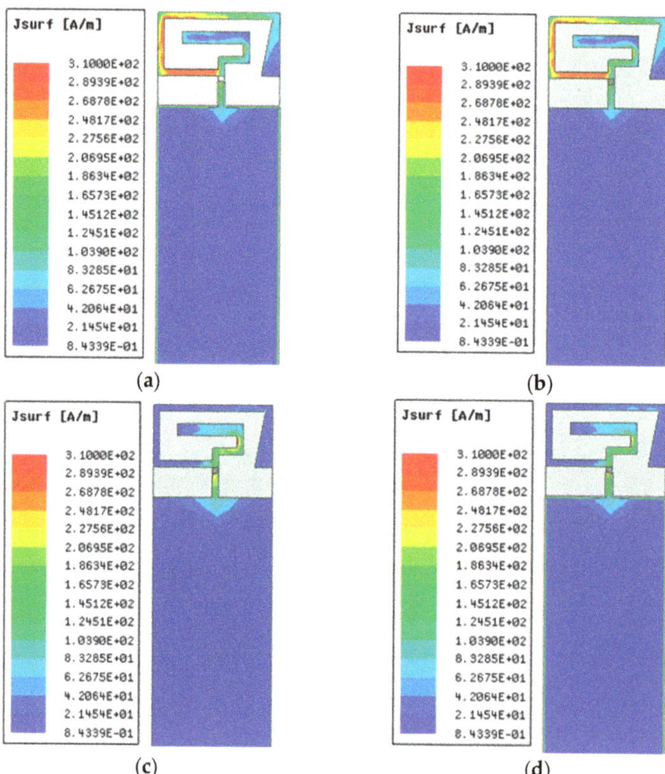

(a)　(b)　(c)　(d)

Figure 2. *Cont.*

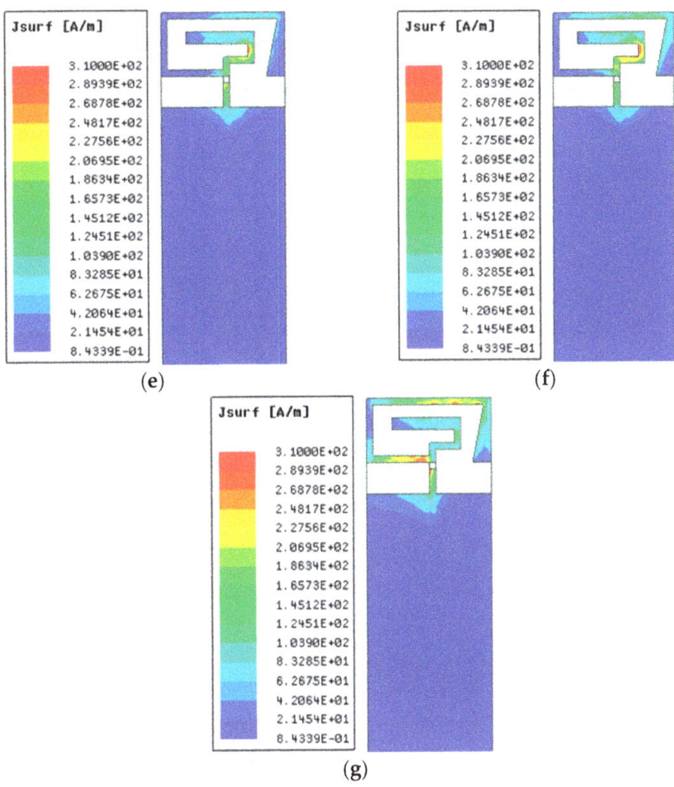

Figure 2. Surface current distribution analysis at: (**a**) 2350 MHz; (**b**) 2450 MHz; (**c**) 5080 MHz; (**d**) 5500 MHz; (**e**) 5890 MHz; (**f**) 6500 MHz; (**g**) 7150 MHz.

3. Validation Analysis, Results, and Discussion

The characteristics and performance of the proposed antenna have been researched in this section. The design analysis process includes the various parameters analysis, the performance with simulation, and measurement results. Figure 3 demonstrates the reflection coefficient's performance with the comparison of simulation and measurement for the proposed dual-band monopole antenna. The validation result agrees with the result simulated and measured. The validation tool employed an EM simulator for simulation results and a vector network analyzer as the equipment of Agilent E7071C for measurement results. The measured impedance bandwidths are defined by 3:1 VSWR, widely used for the internal WWAN antenna design specification [26–30]. The antenna bandwidth defined by VSWR is 3:1 to match the CTIA specification standard with the built-in WWAN antenna design integrated RF active circuit application. Thus, under 6 dB reflection coefficient conduction, the operating frequency can reach the lower band of 2000–2742 MHz and the higher band of 4062–8000 MHz. The bandwidth of the lower band can achieve 31.22%, and the bandwidth of the upper band has an incredible bandwidth of 65.29%. Therefore, the proposed antenna with a dual-band design is suitable for multi-functional wireless communication system standards in autonomous vehicles' communication system devices.

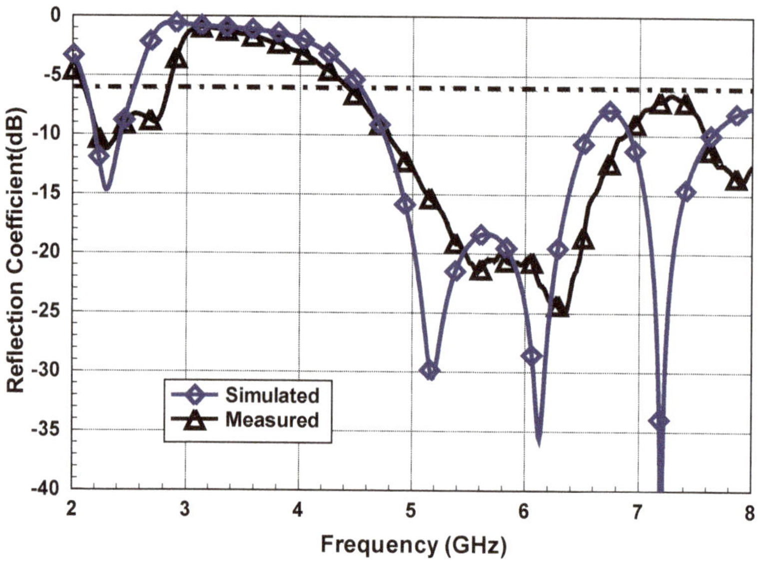

Figure 3. The verification result of the reflection coefficient with simulation and measurement for the proposed monopole antenna.

The progressive analysis of antenna architecture characteristics is shown in Figure 4. Strip 1 has three types to analyze the return loss effect is presented in Figure 4a. The upper operating band is controlled by strip 2 when only strip 2 is without the strip 1 condition as the type 1 structure. Type 2 is with both strip layouts, which can generate the lower and upper operating band. Moreover, the end of strip 1 has been designed with progressive impedance change from narrow to wide, the performance of return loss can achieve a better impedance for the suggestion antenna design as the type 3 antenna configuration is shown in Figure 4a. Discuss that the length change of W9 has a significant influence on the high-frequency response, as shown in Figure 4b. Therefore, the length of W9 equal to 5.5 mm is the best impedance matching response.

The antenna radiation performance verification uses the AMS-8100 model anechoic chamber antenna measurement system manufactured by ETS-Lindgren, as shown the Figure 5.The simulation and measurement radiation patterns are presented in Figure 6, including four operating bands with 2450, 5500, 6500, and 7500 MHz. The radiation results reach omnidirectional modes are good agreement with simulated and measured. Furthermore, the antenna gains and efficiencies are also shown in Figure 7. The results are demonstrated that both performances achieve very close with simulation and measurement. The gain can obtain about 2.7 dBi in the lower band, and the gain can get about 6.75 dBi in the upper band. Thus, the antenna efficiency can reach 91% for the lower band, and the antenna efficiency can achieve 93% for the upper band. In addition, the proposed antenna was also compared with recent literature and listed the bandwidth and dimensions, as the Table 1 The proposed antenna characteristic is shown as compact and widely operating bands. This table uses the journal papers of the past two years for antenna design comparison. The proposed antenna design has a larger bandwidth than the literature [26–30]. The antenna's gain also has a comparable measurement result to the literature [26–30]. The antenna size of the proposed antenna is smaller than in the literature [26–30]. The system ground size is smaller than the literature [26,30]. The overall antenna size is smaller than in the literature [26,28,30].

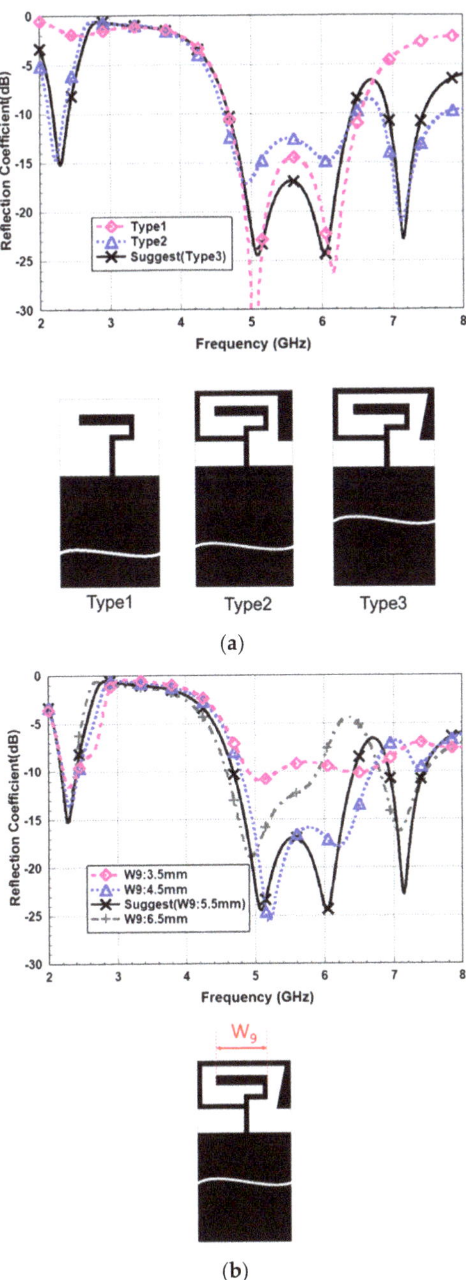

Figure 4. The evolution analysis with various types and lengths for the proposed antenna: (**a**) evolution analysis of antenna design with type 1, 2, and 3; (**b**) strip length W₉.

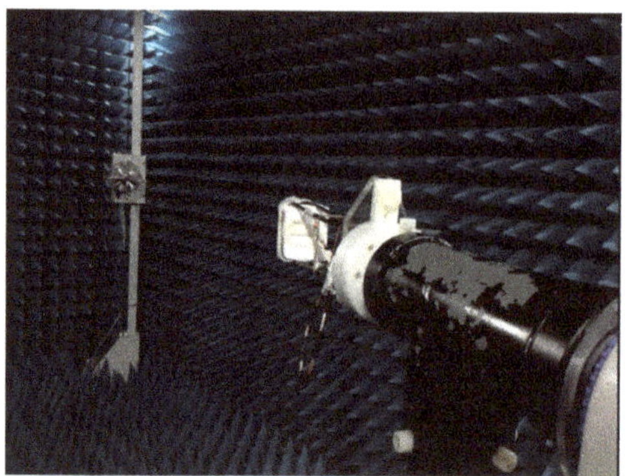

Figure 5. Photographs of the fabricated antenna PCB and the process of measurement.

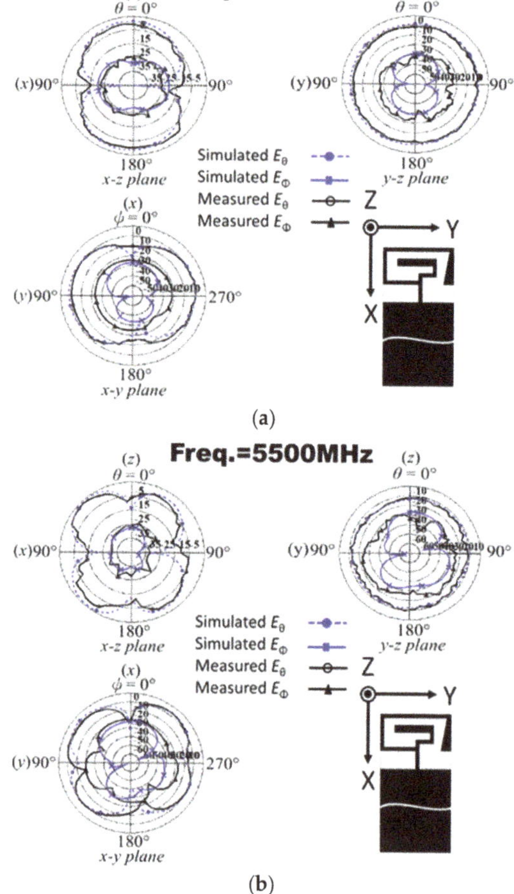

Figure 6. *Cont.*

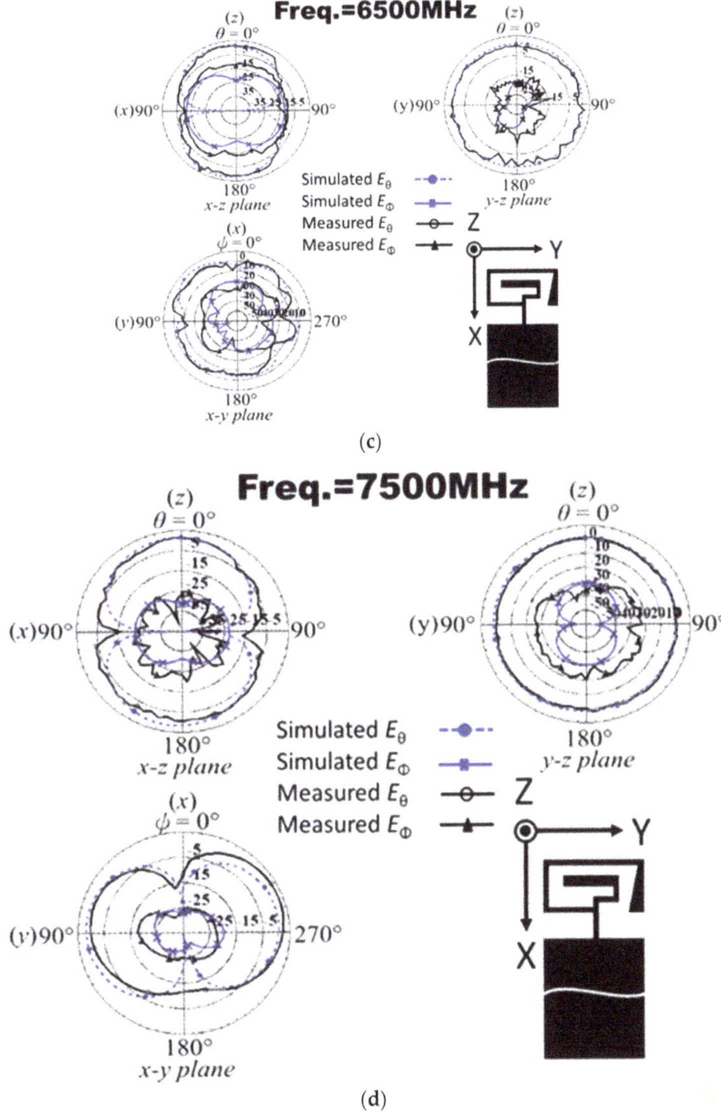

Figure 6. Simulation and measurement of radiation patterns at different operating frequencies: (**a**) 2450 MHz; (**b**) 5500 MHz; (**c**) 6500 MHz; (**d**) 7500 MHz.

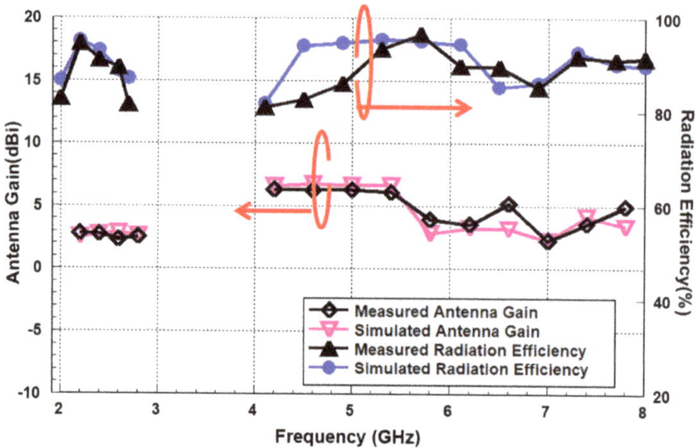

Figure 7. The results of peak gain and radiation efficiency.

This antenna design is suitable for multi-functional wireless communication system standards, covering eight systems. The first wireless communication standard supported is 5G with 13 bands of n1/n7/n30/n38/n40/n41/n46/n47/n53/n79/n90/n95/n96 in two operating frequencies within 2110–2690 MHz and 5150–7125 MHz. The second wireless communication standard system is Beyond-5G for low earth-orbiting satellite (LEO) application, which is operated in X band spectrum and is designed by the International Telecommunication Union (ITU), which is the operating frequency from space to earth in 7250 to 7750 MHz and from earth to space in 7900 to 8400 MHz. The third is the LTE system with fifteen frequency bands, including the support bands of 7/10/16/23/30/34/38/40/41/46/47/65/67/68 in the operating frequency of 2000–2690 MHz and 5150–5920 MHz. The fourth standard system is V2X, and DSRC for autonomous vehicles application belongs to the IEEE Wireless Access in the Vehicular Environment (WAVE), covering the operating frequency between 5850–5925 MHz. The fifth, sixth, and seventh supported system is ISM band in those operating frequencies including 2450–2483.5 MHz, 2300–2690 MHz, 3400–3590 MHz, 5170–5930 MHz, 5150–5350 MHz, and 5725–5850 MHz to correspond to the wireless communication standard with WiFi 2.4G/5G, WiMAX, and Bluetooth. In addition, the WiFi 6 band also is designed in the operating frequency of 5925–7125 MHz. Finally, Sirius/XM Radio has also supported the 2320–2345 MHz band for the in-vehicle infotainment system (IVI system). The results of the proposed antenna have been analyzed by simulation and measurement. As a result, the proposed antenna has stable radiation and a widely broadband characteristic in this research.

Table 1. Comparison of the proposed antenna with other research literature.

References	Publication Year	Overall Size (mm²)	System Ground Size (mm²)	Antenna Size (mm²)	Operation Range (GHz)	Resonant Frequency (GHz)	Bandwidth (%)	Gain (dBi)	Applications
[26]	2020	36 × 38	36 × 21.7	38 × 24.4	4.5–5.57	5.035	30% (1070 MHz)	6	WiFi/LTE/WiMax
[27]	2020	20 × 19	10 × 19	13.5 × 10	2.4–2.54 5.72–5.94	2.74 5.83	5.1% (140 MHz) 3.7% (220 MHz)	2.1 3.5	WBAN
[28]	2020	54 × 37	5 × 54	54 × 32	1.6–2.2	1.9	31.5% (600 MHz)	1.67	-
[29]	2020	24 × 19	8 × 8.5	16 × 16	2.33–2.47 3.68–3.92 5.38–5.75	2.4 6.3 5.565	5.8% (140 MHz) 13.3% (240 MHz) 6.6% (370 MHz)	2.26 3.14 3.73	Bluetooth/5G NR/WLAN
[30]	2021	260 × 208	260 × 200	21 × 8	2.25–2.64 3.25–3.63 5–6.41	2.445 3.44 5.705	5.8% (390 MHz) 13.3% (380 MHz) 6.6% (1410 MHz)	4.38 4 5.68	LTE/WIFI/sub-6 GHz/5G bands
Proposed	-	50 × 10	42.5 × 10	12.5 × 7.5	2–2.74 4.062–8	2.46 6.205	31.22% (740 MHz) 65.29% (3938 MHz)	2.7 6.75	5G, B5G, 4G, V2X, ISM band of WLAN, Bluetooth, WiFi 6 band and WiMAX, Sirius/XM Radio

4. Conclusions

Herein, we have presented the proposed antenna structure with two branches that can achieve dual-band and broadband bandwidth characteristics. Moreover, the antenna performances have been analyzed, validated, and manufactured. Thus, this design is suitable for in-vehicle infotainment and autopilot equipment systems in autonomous vehicle communication systems, including 5G, B5G, 4G, V2X, ISM band of WLAN, Bluetooth, WiFi 6 band, WiMAX, and Sirius/XM Radio application.

Author Contributions: Conceptualization, M.-A.C.; methodology, M.-A.C.; software, M.-A.C.; validation, M.-A.C. and C.-W.Y.; formal analysis, M.-A.C.; investigation, M.-A.C.; resources, M.-A.C.; writing—original draft preparation, M.-A.C.; writing—review and editing, M.-A.C.; visualization, M.-A.C.; supervision, M.-A.C.; project administration, M.-A.C.; funding acquisition, M.-A.C. All authors have read and agreed to the published version of the manuscript.

Funding: This work was supported by the Ministry of Science and Technology, Taiwan, R.O.C under Grant Project MOST 109-2222-E-027-008.

Data Availability Statement: All data are included within manuscript.

Conflicts of Interest: The authors declare no conflict of interest.

References

1. Bassoli, R.; Granelli, F.; Arzo, S.T.; Di Renzo, M. Toward 5G cloud radio access network: An energy and latency perspective. *Trans. Emerg. Telecommun. Technol.* **2021**, *32*, e3669. [CrossRef]
2. Zebari, G.M.; Zebari, D.A.; Al-zebari, A. Fundamentals of 5G Cellular Networks: A Review. *J. Inf. Technol. Inform.* **2021**, *1*, 1–5.
3. Kao, L.-C.; Liao, W. 5G intelligent A+: A pioneer multi-access edge computing solution for 5G private networks. *IEEE Commun. Stand. Mag.* **2021**, *5*, 78–84. [CrossRef]
4. Gao, G.; Li, W. Architecture of visual design creation system based on 5G virtual reality. *Int. J. Commun. Syst.* **2021**, e4750.
5. Alshouiliy, K.; Agrawal, D.P. Confluence of 4g lte, 5g, fog, and cloud computing and understanding security issues. In *Fog/Edge Computing For Security, Privacy, and Applications*; Springer: Cham, Switzerland, 2021; pp. 3–32.
6. Chan, T.K.; Chin, C.S. Review of autonomous intelligent vehicles for urban driving and parking. *Electronics* **2021**, *10*, 1021. [CrossRef]
7. Mahmoud, H.H.H.; Amer, A.A.; Ismail, T. 6G: A comprehensive survey on technologies, applications, challenges, and research problems. *Trans. Emerg. Telecommun. Technol.* **2021**, *32*, e4233.
8. Dao, N.-N.; Pham, Q.-V.; Tu, N.H.; Thanh, T.T.; Bao, V.N.Q.; Lakew, D.S.; Cho, S. Survey on aerial radio access networks: Toward a comprehensive 6g access infrastructure. *IEEE Commun. Surv. Tutor.* **2021**, *23*, 1193–1225. [CrossRef]
9. Guo, F.; Yu, F.R.; Zhang, H.; Li, X.; Ji, H.; Leung, V.C. Enabling massive IoT toward 6G: A comprehensive survey. *IEEE Internet Things J.* **2021**, *8*, 11891–11915. [CrossRef]
10. Kulkarni, J. Wideband cpw-fed oval-shaped monopole antenna for wi-fi5 and wi-fi6 applications. *Prog. Electromagn. Res. C* **2021**, *107*, 173–182. [CrossRef]
11. Kulkarni, J.; Sim, C.Y.D. Multiband, miniaturized, maze shaped antenna with an air-gap for wireless applications. *Int. J. RF Microw. Comput.-Aided Eng.* **2021**, *31*, e22502. [CrossRef]
12. Xu, H.; Chen, Z.; Zhao, Z.; Liu, H.; Zhu, S. A flexible and compact tri-band antenna for vehicular wireless video transmission systems. *Int. J. RF Microw. Comput.-Aided Eng.* **2021**, *31*, e22741. [CrossRef]
13. Wong, K.-L.; Chang, H.-J.; Wang, C.-Y.; Wang, S.-Y. Very-low-profile grounded coplanar waveguide-fed dual-band WLAN slot antenna for on-body antenna application. *IEEE Antennas Wirel. Propag. Lett.* **2019**, *19*, 213–217. [CrossRef]
14. Wen, L.; Gao, S.; Yang, Q.; Luo, Q.; Yin, Y.; Ren, X.; Wu, J. A compact monopole antenna with filtering response for WLAN applications. In Proceedings of the 2019 International Symposium on Antennas and Propagation (ISAP), Xi'an, China, 27–30 October 2019; pp. 1–3.
15. Liu, D.Q.; Zhang, M.; Luo, H.J.; Wen, H.L.; Wang, J. Dual-band platform-free PIFA for 5G MIMO application of mobile devices. *IEEE Trans. Antennas Propag.* **2018**, *66*, 6328–6333. [CrossRef]
16. Ali, T.; Khaleeq, M.M.; Biradar, R.C. A multi-band reconfigurable slot antenna for wireless applications. *AEU-Int. J. Electron. Commun.* **2018**, *84*, 273–280. [CrossRef]
17. Rezapour, M.; Rashed-Mohassel, J.; Keshtkar, A.; Naser-Moghadasi, M. Isolation enhancement of rectangular dielectric resonator antennas using wideband double slit complementary split ring resonators. *Int. J. RF Microw. Comput.-Aided Eng.* **2019**, *29*, e21746. [CrossRef]
18. Ng, W.-H.; Lim, E.-H.; Bong, F.-L.; Chung, B.-K. Folded patch antenna with tunable inductive slots and stubs for UHF tag design. *IEEE Trans. Antennas Propag.* **2018**, *66*, 2799–2806. [CrossRef]
19. Kulkarni, J. Multi-band printed monopole antenna conforming bandwidth requirement of GSM/WLAN/WiMAX standards. *Prog. Electromagn. Res. Lett.* **2020**, *91*, 59–66. [CrossRef]

20. Ghaffar, A.; Li, X.J.; Hussain, N.; Awan, W.A. Flexible frequency and radiation pattern reconfigurable antenna for multi-band applications. In Proceedings of the 2020 4th Australian Microwave Symposium (AMS), Sydney, NSW, Australia, 13–14 February 2020; pp. 1–2.
21. Yang, G.; Zhang, S.; Li, J.; Zhang, Y.; Pedersen, G.F. A multi-band magneto-electric dipole antenna with wide beam-width. *IEEE Access* **2020**, *8*, 68820–68827. [CrossRef]
22. Chung, M.A. Embedded 3D multi-band antenna with ETS process technology covering LTE/WCDMA/ISM band operations in a smart wrist wearable wireless mobile communication device design. *IET Microw. Antennas Propag.* **2020**, *14*, 93–100. [CrossRef]
23. Camacho-Gomez, C.; Sanchez-Montero, R.; Martinez-Villanueva, D.; Lopez-Espi, P.-L.; Salcedo-Sanz, S. Design of a multi-band microstrip textile patch antenna for LTE and 5G services with the CRO-SL ensemble. *Appl. Sci.* **2020**, *10*, 1168. [CrossRef]
24. Kumar, C.M.; Muvvala, N.K. Multi band metamaterial inspired L type slot patch antenna. In Proceedings of the 2020 International Conference on Smart Technologies in Computing, Electrical and Electronics (ICSTCEE), Bengaluru, India, 9–10 October 2020; pp. 34–38.
25. Khan, M.I.; Chandra, A.; Kumar, V.; Das, S. A planar dual band dual polarized slot antenna using coplanar waveguide. In Proceedings of the 2018 IEEE International Students' Conference on Electrical, Electronics and Computer Science (SCEECS), Bhopal, India, 24–25 February 2018; pp. 1–4.
26. Ghouz, H.H.M.; Sree, M.F.A.; Ibrahim, M.A. Novel wideband microstrip monopole antenna designs for WiFi/LTE/WiMax devices. *IEEE Access* **2020**, *8*, 9532–9539. [CrossRef]
27. Le, T.T.; Yun, T.-Y. Miniaturization of a dual-band wearable antenna for WBAN applications. *IEEE Antennas Wirel. Propag. Lett.* **2020**, *19*, 1452–1456. [CrossRef]
28. Zhang, H.; Chen, D.; Zhao, C. A novel printed monopole antenna with folded stepped impedance resonator loading. *IEEE Access* **2020**, *8*, 146831–146837. [CrossRef]
29. Sreelakshmi, K.; Rao, G.S.; Kumar, M. A compact grounded asymmetric coplanar strip-fed flexible multi-band reconfigurable antenna for wireless applications. *IEEE Access* **2020**, *8*, 194497–194507. [CrossRef]
30. Kulkarni, J. Multiband triple folding monopole antenna for wireless applications in the laptop computers. *Int. J. Commun. Syst.* **2021**, *34*, e4776. [CrossRef]

 *electronics*

Article

4 × 4 MIMO Antenna System for Smart Eyewear in Wi-Fi 5G and Wi-Fi 6e Wireless Communication Applications

Ming-An Chung *, Cheng-Wei Hsiao, Chih-Wei Yang and Bing-Ruei Chuang

Department of Electronic Engineering, National Taipei University of Technology, Taipei 10608, Taiwan; t109368536@ntut.org.tw (C.-W.H.); t109368113@ntut.edu.tw (C.-W.Y.); t109368056@ntut.org.tw (B.-R.C.)
* Correspondence: minganchung@ntut.edu.tw

Abstract: This paper proposes a small-slot antenna system (50 mm × 9 mm × 2.7 mm) for 4 × 4 multiple-input multiple-output (MIMO) on smart glasses devices. The antenna is set on the plastic temple, and the inverted F antenna radiates through the slot in the ground plane of the sputtered copper layer outside the temple. Two symmetrical antennas and slots on the same temple and series capacitive elements enhance the isolation between the two antenna ports. When both temples are equipped with the proposed antennas, 4 × 4 MIMO transmission can be achieved. The antenna substrate is made of polycarbonate (PC), and its thickness is 2.7 mm ($\varepsilon r = 2.85$, $\tan \delta = 0.0092$). According to the actual measurement results, this antenna has two working frequency bands when the reflection coefficient is lower than −10dB, its working frequency bandwidth at 4.58–5.72 GHz and 6.38–7.0 GHz. The proposed antenna has a peak gain of 4.3 dBi and antenna efficiency of 85.69% at 5.14 GHz. In addition, it also can obtain a peak gain of 3.3 dBi and antenna efficiency of 82.78% at 6.8 GHz. The measurement results show that this antenna has good performance, allowing future smart eyewear devices to be applied to Wi-Fi 5G (5.18–5.85 GHz) and Wi-Fi 6e (5.925–7.125 GHz).

Keywords: smart eyewear antenna; slot antenna; MIMO system; Wi-Fi 6e; polycarbonate

Citation: Chung, M.-A.; Hsiao, C.-W.; Yang, C.-W.; Chuang, B.-R. 4 × 4 MIMO Antenna System for Smart Eyewear in Wi-Fi 5G and Wi-Fi 6e Wireless Communication Applications. *Electronics* **2021**, *10*, 2936. https://doi.org/10.3390/electronics10232936

Academic Editors: Issa Tamer Elfergani, Raed A. Abd-Alhameed and Abubakar Sadiq Hussaini

Received: 25 October 2021
Accepted: 24 November 2021
Published: 26 November 2021

Publisher's Note: MDPI stays neutral with regard to jurisdictional claims in published maps and institutional affiliations.

1. Introduction

Smart wearable devices that are compact, lightweight, and ergonomic are quickly becoming mainstream. With the development of communication technology, smart glasses have gradually received widespread attention, such as the Google smart glasses in 2012, smart goggles Form Swim Goggles in 2019, and the recent Facebook smart glasses in 2021. The antenna design specifications of smart glasses mostly use Bluetooth and Wi-Fi wireless communication for indoor communication and entertainment. Moreover, they communicate with mobile phones and wearable devices via Bluetooth [1–3]. However, with the rapid development of smart glasses research, smart glasses devices will no longer only use Bluetooth or Wi-Fi antenna technology, such as a WiGig array module antenna design on the frame [4] or LTE multi-band antenna design on the temple [5].

At present, there are various types of antennas in smart wearable devices and smartphones. These devices usually demand ultra-wideband (UWB) operation. Due to the limited antenna design space, slot antennas or microstrip antennas are the best choice for these work environments [6–12]. In recent years, articles on the configuration of slot antennas for wireless communication products have emerged. For example, a notebook device uses the shaft block to design the slot antenna to produce a 2 × 2 MIMO antenna system for low frequencies of 700–900 MHz and high frequencies of 1.7–2.6 GHz [7]. In [8], the smartphone uses the coupling effect between the inverted-F antenna and the slot to design a 2 × 2 LTE low-frequency antenna and a 4 × 4 LTE high-frequency antenna in a tiny space. Finally, [10] proposes a smartwatch with a 2 × 2 MIMO antenna system in a metal frame to match the frequency band to achieve the required coverage according to the slot antenna and circuit components.

In the fourth-generation (4G) mobile networks, 2 × 2 MIMO antenna systems have been widely used in various wireless communication products [13]. However, for fifth-generation (5G) mobile communications and Wi-Fi 6 wireless indoor communications, more antennas are needed to support the ultra-high broadband transmission rate in the ultra-high-frequency band and better transmission path reliability. Moreover, it will be a major challenge for antenna researchers to set up a MIMO antenna system with high isolation in space-constrained smart wearable devices, and an asymmetrical antenna architecture can compensate for the problem of poor isolation [14–18].

The antenna of smart glasses is most often placed on the temples. When the antenna is closer to the human body, a lower specific absorption rate (SAR) is required to comply with the regulations of various countries [19,20]. The SAR value is limited to 1 g 1.6 W/kg under the standard established by FCC KDB 248227 in the United States [21], while Europe and the Institute of Electrical and Electronics Engineers (IEEE) require the SAR value to be limited to 10 g 2 W/kg [22]. Because the Wi-Fi 6e working frequency is greater than 6 GHz, FCC needs to measure the power density (PD), and its limit value is 10 W/m^2 [23].

In the current research, articles on the application of Wi-Fi 6e frequency to smart glasses have not yet been published. Therefore, this paper proposes a Wi-Fi 5G and Wi-Fi 6e dual-band 4 × 4 MIMO antenna system for plastic glasses. In this article, Section 2 describes the design process of the glasses antenna and simulates the slot size and circuit matching. Section 3 observes the impact caused by the antenna approaching the head and the simulation of the head SAR value. In Section 4, we describe a physical glasses antenna based on the previous simulation results and provide measurements of the antenna gain, antenna efficiency, and 2D radiation pattern; we then provide a comparison table with other references. Finally, this paper is concluded in Section 5.

2. Antenna Design

Figure 1 shows the 4 × 4 MIMO slot antenna system applied to the Wi-Fi 6e band proposed in this paper. The antenna system can be divided into an inverted-F antenna and a metal ground plane. The inverted-F antenna is manufactured by Laser Direct Structuring (LDS) on the inner side of the temple with a substrate made of polycarbonate (PC). The substrate thickness is 2.7 mm, the dielectric constant of polycarbonate is 2.85, and the dielectric loss tangent is 0.0092. The outside of the PC temples uses a sputtering coating process to ensure that the antenna's metal ground and reserve slots have a coupling effect.

It can be seen from Figure 1a that the glasses proposed in this paper refer to the currently popular plastic glasses, and the overall size is 155 mm × 145 mm × 50 mm. The left temple is equipped with antenna 1 and antenna 2, and the right temple is equipped with antenna 3 and antenna 4. The left leg antenna and the right leg antenna have a symmetrical structure. Figure 1b is a detailed size figure of a single temple antenna. For the size code in the figure, please refer to Table 1. The slot antenna occupies an area of 50 mm × 9 mm × 2.7 mm on the temple. The slot extends from the outside of the temple top, the length of the SL is 5.5 mm, and the width of the SW is 2 mm. The inverted-F antenna feeds from the ground side at 6 mm, and there is a 1 pF capacitor in series in the middle of the antenna. The length L of the antenna is 6 mm, and it reaches the ground through the Via hole. Therefore, the four antennas installed on the plastic glasses are not obtrusive and can meet the requirements of the metal frame of the smart wearable device.

2.1. Slot Antenna Simulation and Analysis

The antenna design steps proposed in this paper are shown in Figure 2. In Figure 2a, an inverted-F antenna that can be coupled to the slot is shown, which is grounded through a Via hole with a diameter of 1 mm. The ground plane is a metal copper layer covering the outer, top, and bottom sides of the glasses, and, according to the simulation results, slot holes are present on the outside grounding surface. Because the temple area is large, as Figure 2b shows, the eyewear device has another symmetrical antenna at a distance of 18 mm from the inverted-F antenna. Figure 2c shows two symmetrical inverted-F antennas with series capacitors, allowing the antenna to generate the resonance frequency of the target operating frequency band and, at the same time, enhance the isolation between the two antennas. Isolation performance is essential for MIMO antennas.

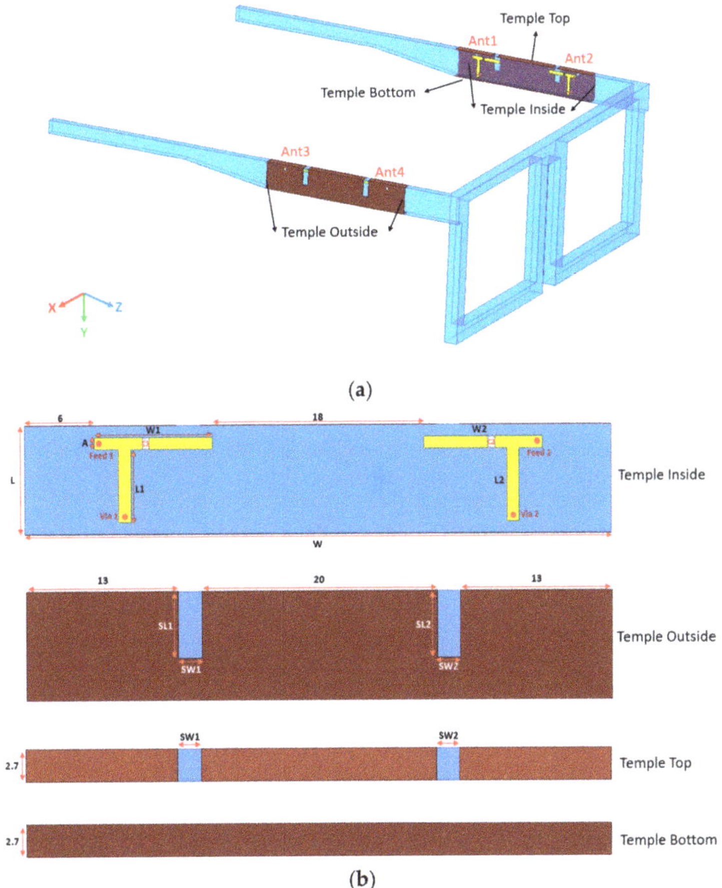

Figure 1. (**a**) Geometry of the proposed 4 × 4 MIMO glasses antennas; (**b**) the dimensions of inverted-F antenna and slotted copper ground plane.

Table 1. The dimensions of the proposed antenna.

Parameter	W	L	W1	W2	L1	L2	SL1
Units(mm)	50	9	10	10	6	6	5.5
Parameter	SL2	SW1	SW2	A	Capacitor	C1	C2
Units(mm)	5.5	2	2	1	Units(pF)	1	1

Next, the simulation software High-Frequency Structure Simulator (HFSS) version 2021R1 is used. First, we simulate and analyze the capacitance change, and then simulate and compare the length and width of the slot. The simulation results in Figure 3a show that when the antenna is as in Figure 2a, without series capacitors, the resonance frequency is generated by the entire slot antenna on 4.6 GHz and 6.7 GHz. However, the simulation results cannot cover the Wi-Fi 6e frequency band completely because the overall working frequency band cannot be continuous. Thus, we try to connect the inverted-F antenna in series with capacitive elements and adjust the capacitance value to change the resonance frequency of the slot antenna. When C1 = 0.5 pF, the working bandwidth cannot cover Wi-Fi 5G and Wi-Fi 6e. When C1 = 1 pF, the resonant points of the slot antenna can be obtained at 5.3 GHz and 6.6 GHz, and the working bandwidth can cover the entire Wi-Fi 5G and Wi-Fi 6e. Figure 3b presents a simulation of the presence or absence of a series capacitor. It is found that the series capacitor can improve the isolation between the two antennas, and the S21 is lower than −10 dB.

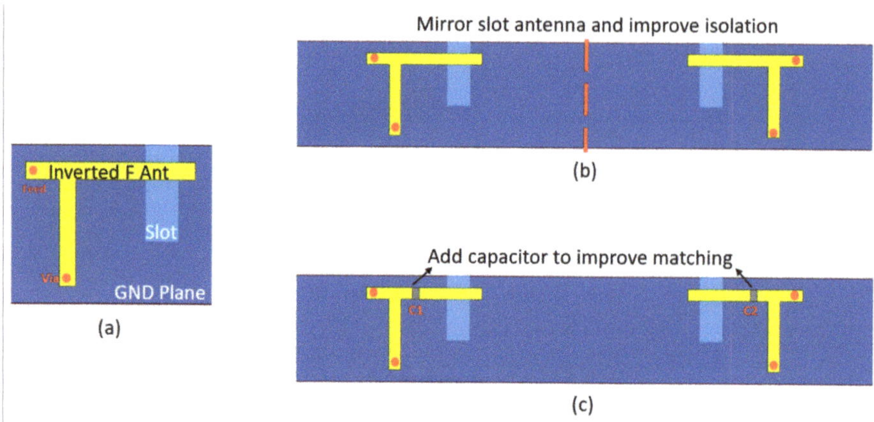

Figure 2. The proposed slot antenna design steps. (**a**) Single slot antenna system; (**b**) Mirrored slot antenna system; (**c**) Series capacitor to improve impedance matching.

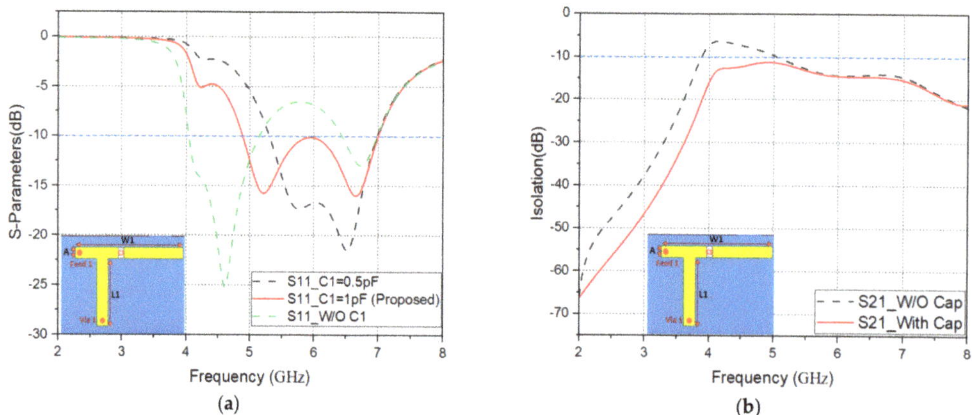

Figure 3. S-parameter simulation results as: (**a**) capacitance change; (**b**) the effect of series capacitance on isolation.

In addition, we also simulate the change of slot size, and the result is shown in Figure 4a. When the slot length SL is 4.5 mm, from the reflection coefficient $|S11| \leq -10$ dB, it can be seen that the resonance frequency will move to the high-frequency direction at 5.3 GHz to 6.6 GHz, and the resonance intensity will be weakened. When the slot length SL is 6.5 mm, the resonance frequency of 6.6 GHz will move to the low-frequency direction, and the working bandwidth will also be reduced. Figure 4b shows that when the slot width SW is 1 mm, the resonance frequency will move to a low frequency at 5.3 GHz. When the slot width is 3 mm, the resonance frequency of 5.3 GHz will move to 6 GHz, and the overall operating frequency band will be weakened. In order to achieve the expected working frequency band and resonance strength, the proposed antenna SL is 5.5 mm, and the SW is 2 mm.

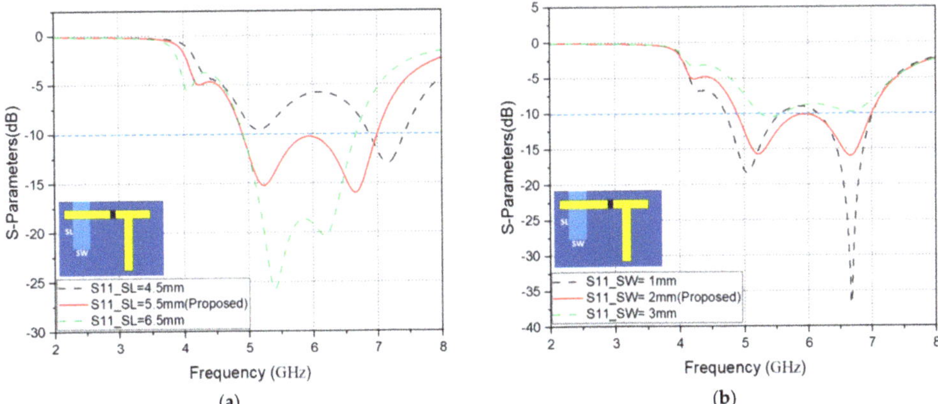

Figure 4. S-parameter simulation results as: (**a**) slot length SL changes; (**b**) slot length SW changes.

2.2. Surface Current Distributions

The operational mode of the slot antenna is based on the current entering from the feed point and exists in the entire inverted-F antenna. At the same time, this current will also be coupled around the slot and allow electromagnetic waves to radiate out of the slot. By adjusting the size of the slot, the target resonance frequency can be within the expected range. The surface current simulation results are shown in Figure 5. When the current enters from the feed point, the proposed antenna will exhibit current coupling on the slot at the expected working frequency band, and it can be found that the working resonance frequency is 5.3 GHz and 6.6 GHz, and the energy is concentrated around the slot on the outside of the temple. From the current energy distribution, it is known that there is almost no current energy distribution between the two antennas, so the symmetrical antenna can provide good isolation.

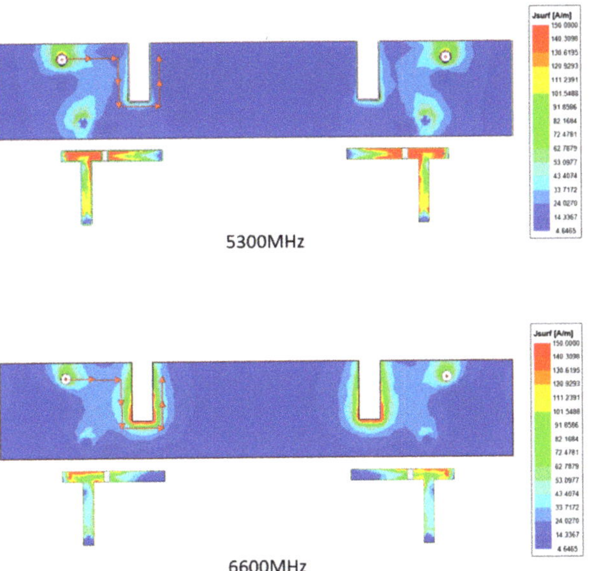

Figure 5. Current distribution paths at 5.3 GHz and 6.6 GHz of Ant1 and Ant2.

3. SAR Simulation and Analysis

The human body is a high-loss material for the antenna. Therefore, the proposed antenna needs to be hung on the human head model for simulation, and we then observe whether the antenna's performance is affected, so that the proposed antenna is more likely to be used in daily life. Figure 6 shows the use of HFSS to simulate the Wi-Fi 5G and Wi-Fi 6e eyewear antenna system, which is placed on a human head model.

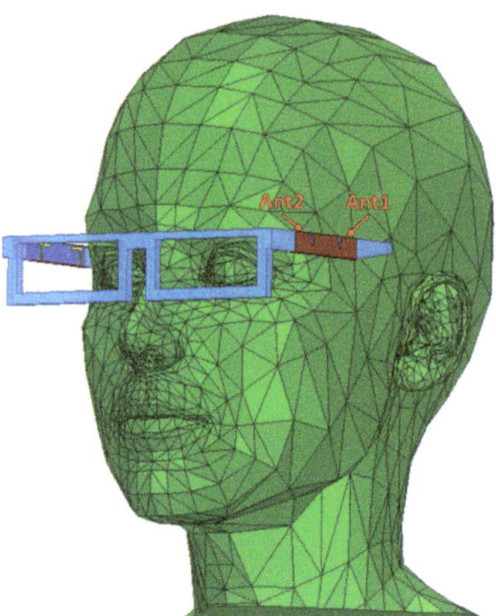

Figure 6. The proposed eyewear antenna on a head model.

Next, we discuss the changes in reflection coefficient and isolation when the proposed antenna is close to the head. It can be seen from Figure 6 that antenna 1 is closer to the head, and antenna 2 is farther from the head. Figure 7 shows the simulation by placing the glasses on the head. However, the proposed antenna is symmetrical, so only antenna 1 and antenna 2 are simulated. Figure 7a shows that when antenna 1 is at the head position, the working bandwidth is shortened due to the influence of the human body, and the reflection coefficient of antenna 2 is not significantly different between the head and the free space. Wireless products on the market can be accepted if the reflection coefficient is lower than −6 dB [24–26], and when the proposed antenna worn on the head that the S-parameter is lower than −6 dB. Therefore, it matches the CTIA specification standard. Figure 7b shows the isolation between antenna 1 and antenna 2, and the isolation is almost lower than −10 dB when hung on the head or in free space.

Smart wearable devices are the electronic products closest to the human body; these electronic products will transmit wireless communication through antennas, and electromagnetic radiation will definitely be generated during the transmission. Therefore, laws and regulations of various countries all have Specific Absorption Rate (SAR) specifications, which are mainly used to limit the electromagnetic radiation absorption caused by electronic products to the human body. For example, the FCC safety limit is 1 g SAR 1.6W/kg, and the European CE safety limit is 10 g SAR 2 W/kg. The SAR calculation method is the radiant power energy absorbed by a unit volume in a specific time, as shown in the following equation [22]:

$$SAR = \frac{d}{dt}\left(\frac{dW}{dm}\right) = \frac{d}{dt}\left(\frac{dW}{\rho dV}\right) \tag{1}$$

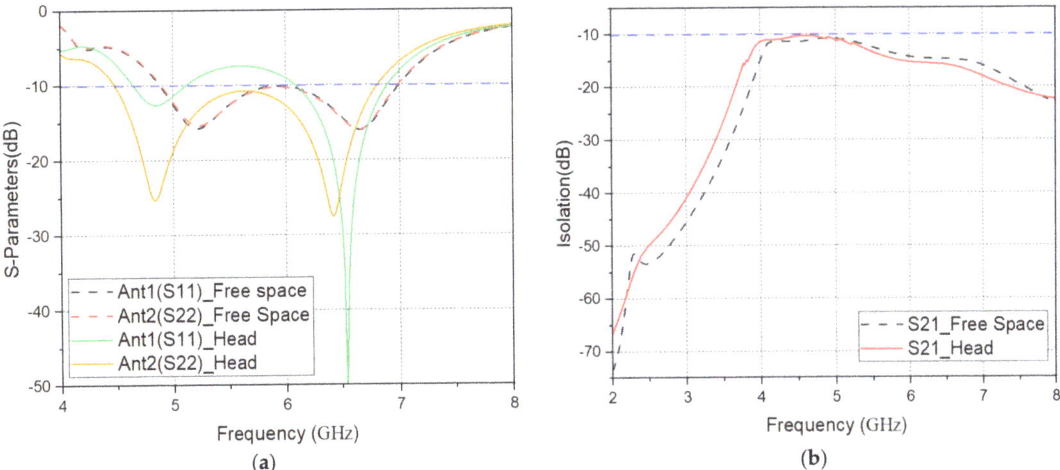

Figure 7. Simulation in head and free space: (**a**) the S-parameters of Ant.1 and Ant.2; (**b**) the isolation between Ant.1 and Ant.2.

Antenna 1 and antenna 3 are closest to the human head in the proposed 4×4 MIMO slot antenna system, and the antenna structure is symmetrical, so only the SAR value of antenna 1 is simulated, and the SAR simulation software uses Sim4Life to operate. As shown in Figure 8, when antenna 1 is 10 mm away from the head, the input power from the feed point is 0.1 W (20 dBm), and the frequency at 5.3 GHz and 6.6 GHz to calculate the values of 1 g SAR and 10 g SAR. According to the simulation results, the 1 g SAR and 10 g SAR of antenna 1 at 5.3 GHz are 1.39 W/kg and 0.722 W/kg, respectively, and the 1 g SAR and 10 g SAR at 6.6 GHz are 0.938 W/kg and 0.467 W/kg, respectively. Thus, the above SAR values are all lower than the certification limit.

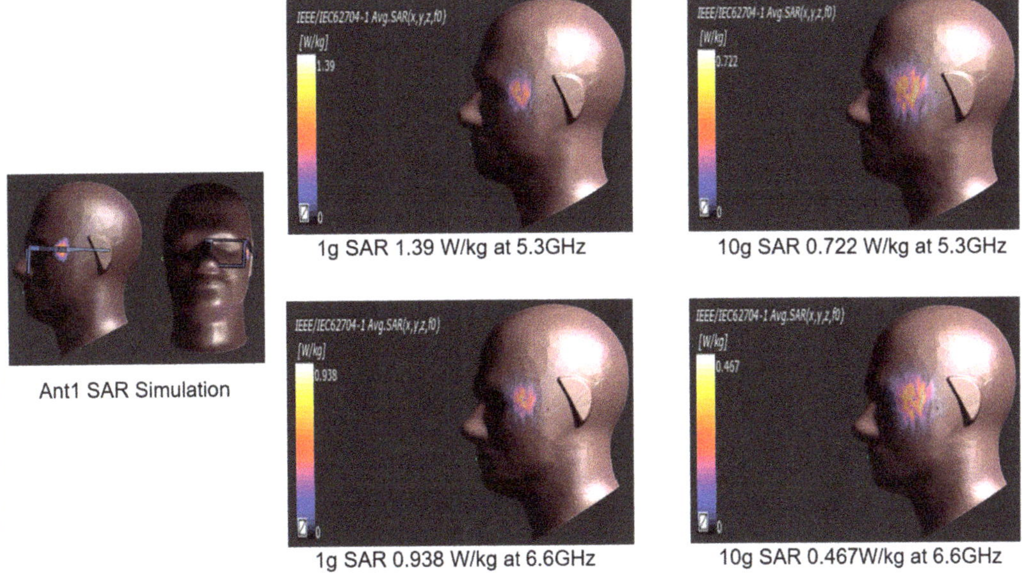

Figure 8. Ant.1 SAR simulation results at 5.3 GHz and 6.6 GHz.

According to the FCC's regulatory test requirements, when the wearable device supports the frequency band above 6 GHz, the power density (PD) needs to be measured. PD test frequency is selected which the frequency point has the worst SAR value, and PD safety limit value is 10 W/m^2. The equation is as follows [27]:

$$S_{inc}^{m}(A, d, \beta_\chi) = \frac{1}{A} \int_{A} |Re[E(d, \beta_\chi) \times H^*(d, \beta_\chi)]| d_s \tag{2}$$

Figure 9 shows antenna 1 when the input power from the feed point is 0.1 W (20 dBm), and the frequency is 6.6 GHz to calculate the PD value. According to the simulation results, the PD of antenna 1 at 6.6 GHz is 6.64 W/m^2, which is lower than the limit of 10 W/m^2.

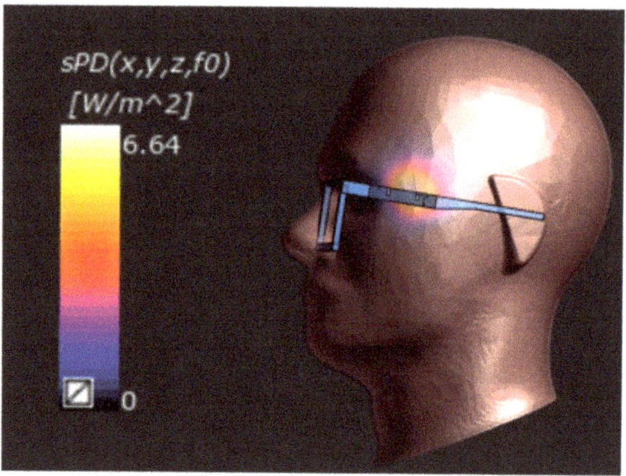

PD 6.64W/m^2 at 6.6GHz

Figure 9. Ant.1 SAR simulation results at 5.3 GHz and 6.6 GHz.

4. Experimental Characterization

The actual work of the proposed 4 × 4 MIMO eyewear antenna has been completed, as shown in Figure 10. According to the above simulation results, the inverted-F antenna on the inside of the temples needs to reserve a block for soldering series capacitors. Unfortunately, the antenna substrate is made of PC (PC's melting point is 220 °C), which cannot withstand general soldering operations. Therefore, we use low-temperature soldering to prevent abnormal antenna performance due to the deformation of the temples. Figure 10a shows a photo of the 4 × 4 MIMO plastic glasses. Figure 10b shows a ground plane of the sputtered copper layer with slots on the outside of the temple. Figure 10c shows an inverted-F antenna created by the LDS process on the inside of the temple.

4.1. S-Parameter Analysis

Because the proposed glasses' antenna substrate is thick and the material is plastic with a low melting point, it is not suitable to use a copper pipe or SMA connector for measurement. Therefore, we use a coaxial cable with low-temperature soldering for measurement. The vector network analyzer model is Agilent E5071C, and we use the SOLT method to calibrate before the measurement, but deviation is present due to soldering and the inability to calibrate the coaxial cable. Therefore, the S-parameter measurement setup in free space is shown in Figure 11a, and the glasses are worn on the human head to measure the S-parameters, as shown in Figure 11b.

Figure 12a compares the S-parameter measurement and simulation of the single temple antenna 1. Due to the soldering and fabrication tolerances, frequency resonance may be weakened, but the measurement result is very similar to the simulation. Figure 12b compares the measurement and simulation of the isolation between antenna 1 and antenna 2.

The isolation between the measurement and simulation can be lower than −10 dB, complying with the MIMO antenna system's basic specifications. Figure 12c is the S-parameter measurement of four antennas. It is found that the effective working frequency bands of the four antennas are all in line with expectations. However, the reflection coefficient of antenna 4 is different from those of the other antennas. This deviation may be due to the soldering and fabrication tolerances.

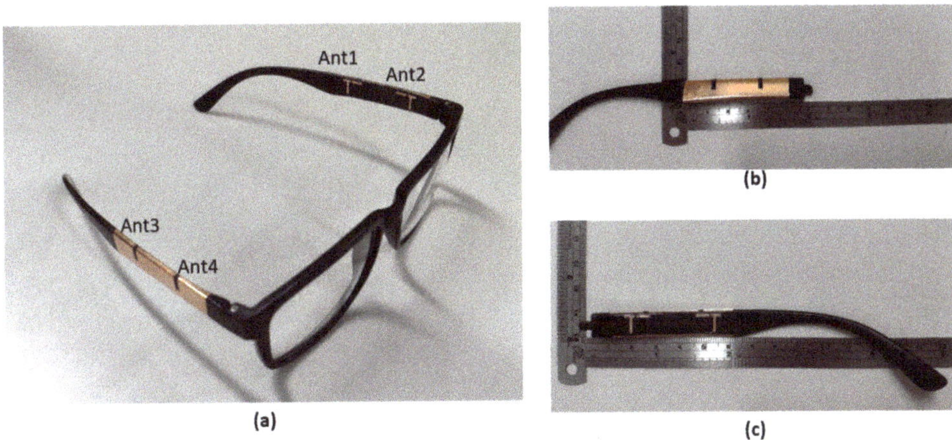

Figure 10. Photographs of the proposed eyewear antennas. (**a**) 4 × 4 MIMO plastic glasses; (**b**) Outside of the temple; (**c**) Inside of the temple.

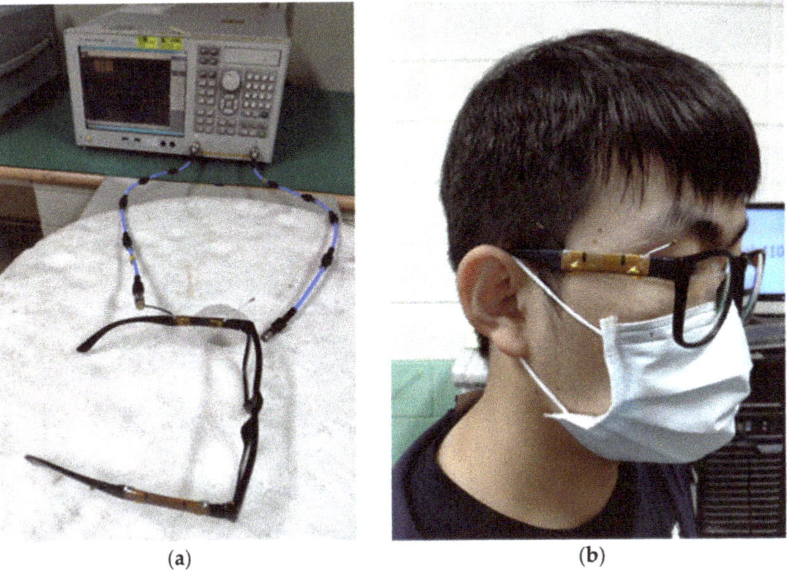

Figure 11. Photographs of (**a**) the measurement setup for the S-parameter; (**b**) glasses on the human head to measure S-parameters.

The proposed antenna is designed on the glasses device, so it is necessary to measure the glasses' S-parameter on the human head. Because the proposed antenna is a symmetric antenna, we use the left-side temple with antenna 1 and antenna 2 to measure the

S-parameter. Figure 13a shows the reflection coefficient measurement and simulation of antenna 1 and antenna 2. The simulation and measurement results are very similar. For antenna 2, simulation and measurement both move towards low frequencies. Figure 13b shows the isolation result of antenna 1 and antenna 2. Both the simulation and measurement yield values lower than −10 dB, and the trend is consistent.

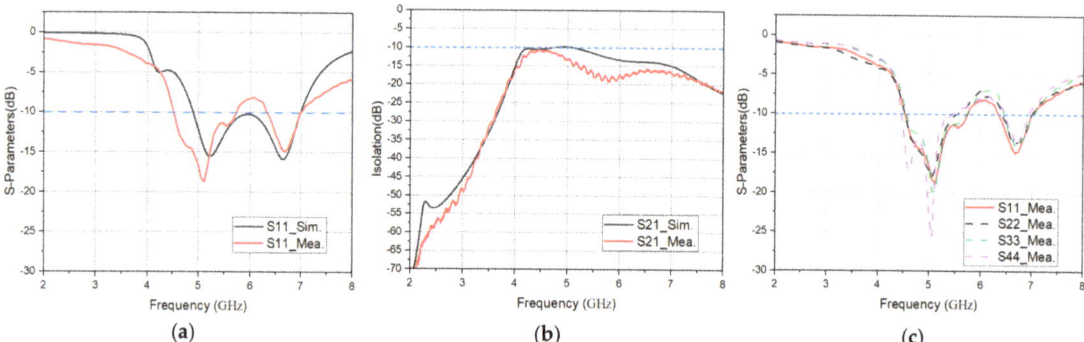

Figure 12. Measured result for the proposed antenna: (**a**) Ant.1 reflection coefficient; (**b**) Ant.1 and Ant.2 isolation; (**c**) Ant.1~Ant.4 reflection coefficient.

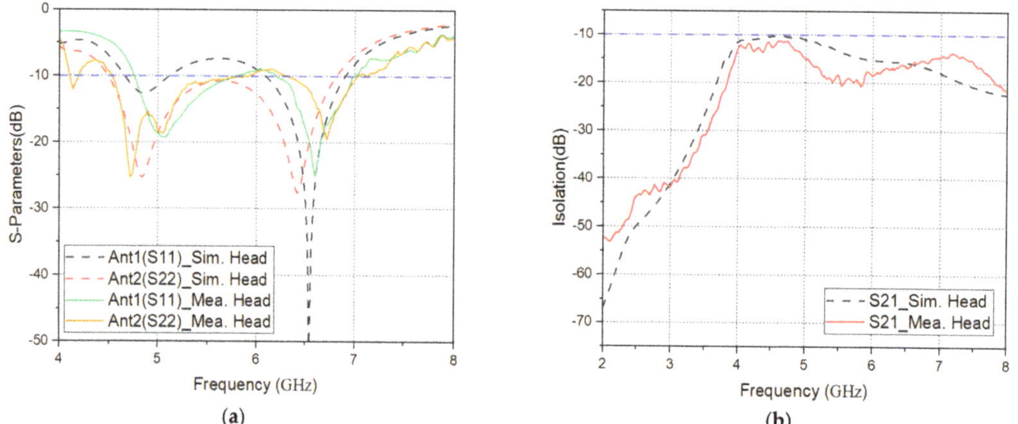

Figure 13. Measurement results for glasses on the human head: (**a**) Ant.1 and Ant.2 reflection coefficient; (**b**) Ant.1 and Ant.2 isolation.

4.2. Radiation Pattern

From the results in Figure 12c, it can be seen that the resonance modes of the four antennas are similar. Therefore, antenna 1 and antenna 2 are used for the following far-field test in this paper. The measurement field used this time is a standard far-field chamber, the network analyzer is the model Anritsu MS46524B, and the measurement frequency band of the horn antenna in the chamber is 0.1–8 GHz. Figure 14 shows a photo of the test environment of antenna 1 in the far-field chamber, and the XYZ axis of the simulation software corresponds to the XYZ axis of the chamber for elevation.

The proposed glasses' antenna works in linear polarization. In a specific receiving range, it will have better signal reception than a circularly polarized antenna. Currently, linearly polarized antennas are widely used in wireless communication products. Figures 15 and 16 show the 2D gain field patterns of the simulated and measured eyewear antenna in the XY plane, the XZ plane, and the YZ plane. From Figures 15 and 16, it can be seen that the simulated and measured radiation modes have the same trend when antenna 1 and antenna 2 are at the working frequency of 5.1 GHz and 6.8 GHz.

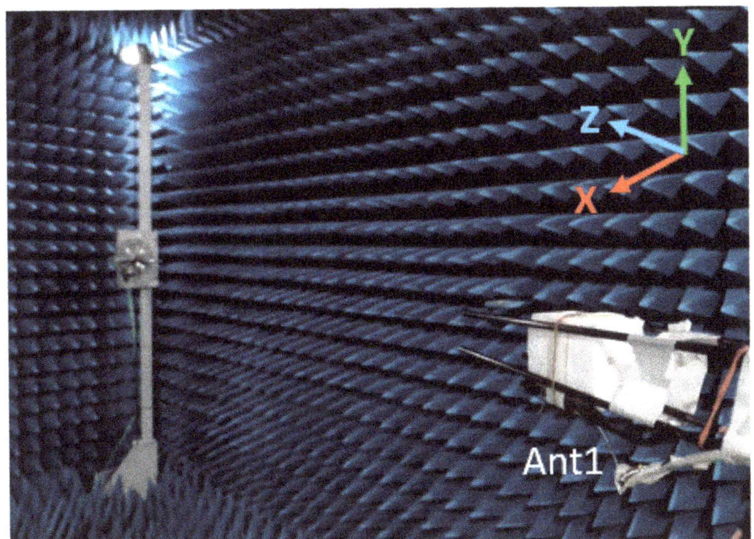

Figure 14. Photograph of Ant.1 erected in a far-field chamber.

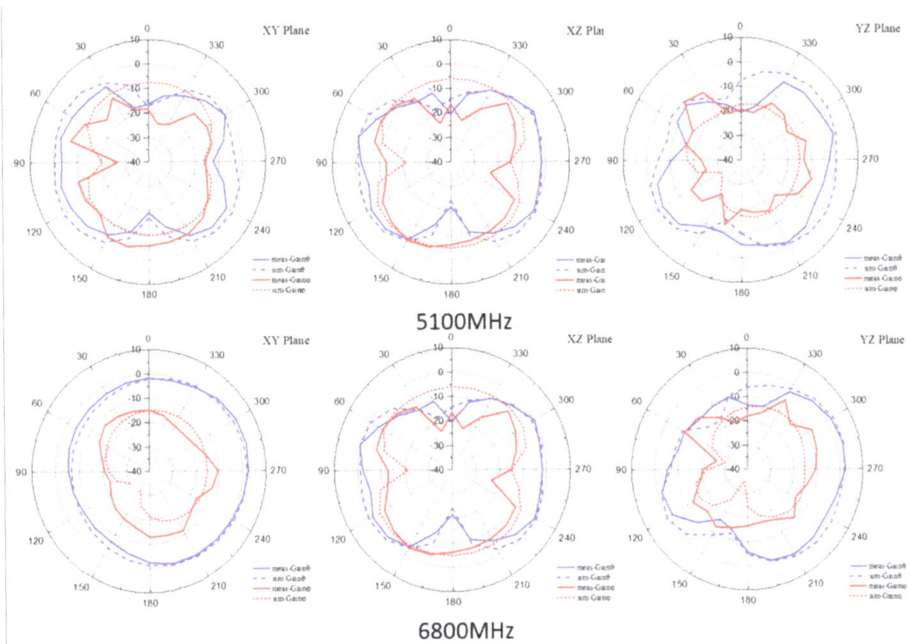

Figure 15. Measured 2D radiation patterns of the proposed eyewear Ant.1.

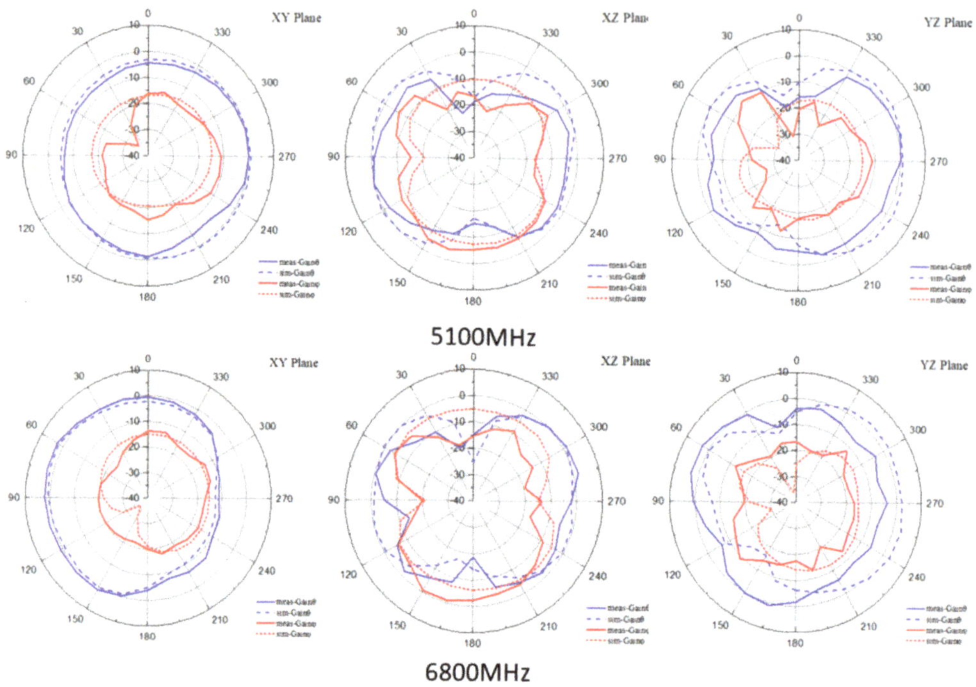

Figure 16. Measured 2D radiation patterns of the proposed eyewear Ant.2.

4.3. Antenna Gain, Efficiency, and ECC

The gain and efficiency of antenna 1 can be seen in Figure 17a,b. The working frequency of antenna 1 can reach 4.3 dBi peak gain and 85.69% antenna efficiency in free space at 5.14 GHz and reach 3.3 dBi and 82.78% antenna efficiency in free space at 6.8 GHz.

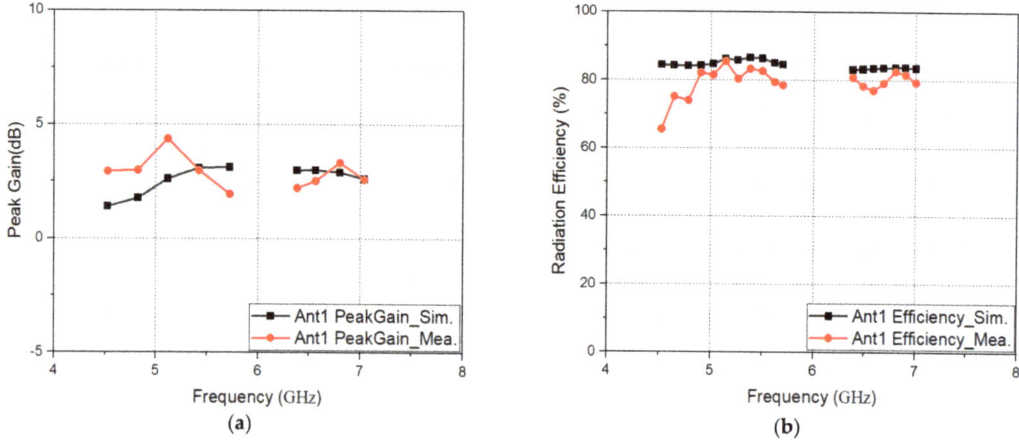

Figure 17. Measured result for the proposed eyewear Ant.1: (**a**) peak gain, (**b**) radiation efficiency.

When the antenna belongs to the MIMO system, the Envelope Correlation Coefficient (ECC) value is usually used to judge whether the antenna's independence is good. The coupling effect between two antennas is poor when the ECC value is lower, and less of the transmission rate will be affected. The internationally recognized ECC limit value is 0.5. The ECC values of the two antennas can be calculated using the S-parameter equation as follows [16]:

$$ECC = \frac{\left|S_{11}^* S_{12} + S_{21}^* S_{22}\right|^2}{\left(1 - |S_{11}|^2 - |S_{21}|^2\right)\left(1 - |S_{22}|^2 - |S_{12}|^2\right)}.$$ (3)

Another ECC calculation method is to use the antenna radiation direction calculation. The ECC value equation of the two antennas is as follows [17]:

$$ECC = \rho_{12} = \frac{\left|\iint_{4\pi}\left[\vec{F}_1(\theta, \varphi) \cdot \vec{F}_2(\theta, \varphi)\right]d\Omega\right|^2}{\iint_{4\pi}\left|\vec{F}_1(\theta, \varphi)\right|^2 d\Omega \cdot \iint_{4\pi}\left|\vec{F}_2(\theta, \varphi)\right|^2 d\Omega}.$$ (4)

In the MIMO antenna system, when port 1 is excited and other ports are connected to a 50-ohm load, the far-field radiation direction generated by antenna 1 can be represented by $\vec{F}_1(\theta, \varphi)$. When port 2 is excited, the far-field radiation direction generated by antenna 2 can be represented by $\vec{F}_2(\theta, \varphi)$.

Figure 18 shows the ECC measurement and simulation comparison of antenna 1 and antenna 2 on a single temple. Again, it can be seen that in the effective working frequency band, the measured and simulated ECC values do not exceed 0.05, and the antenna's independence is good.

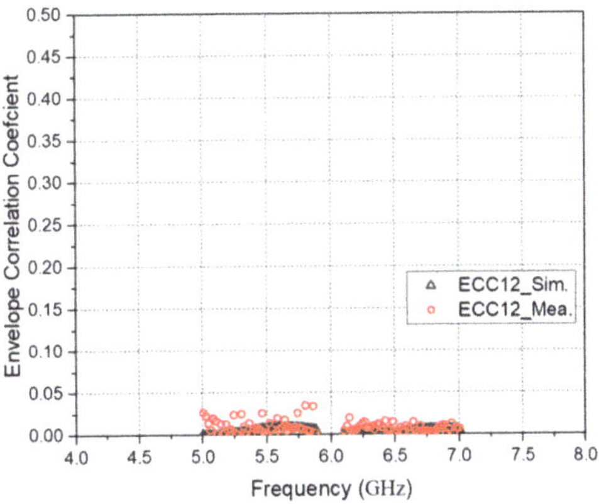

Figure 18. Measured and simulated ECC for Ant.1 and Ant.2.

Table 2 offers a comparison between the proposed antenna and other eyewear antennas. It can be seen that the reference antennas [1,3,14] only support Wi-Fi 2.4 G, and the antenna efficiency is not good. For example, in [3], the antenna efficiency is only 10.7%. The application of 60GHz eyewear antenna is mentioned in references [4] and [28]. Although references [4] and [28] have high antenna gain, but the complex design of the array antenna makes it difficult to realize the MIMO system on the glasses. After comparing the references of LTE frequency bands that references [20] and [29] support multiple LTE frequency bands, but the antenna gain is not good [20] and the operating frequency is not continuous [29]. Reference [30] used an eyewear frame antenna and supported IoT 5.8 G, but it did not possess a MIMO function. In summary, no antenna design supports Wi-Fi 6e in the current references. However, our proposed eyewear antenna has a wide working bandwidth, high antenna efficiency, high antenna gain, supports 4×4 MIMO, and can be applied to the Wi-Fi 5G and Wi-Fi 6e bands.

Table 2. Comparison between proposed work and references.

Ref.	Ant. Type	Apply To	Operating Band/BW (GHz)	Efficiency (%)	Gain (dBi)	MIMO
[1]	IAI	Eyewear	Wi-Fi 2.4 G 2.4–2.5	2.4 G: 46.8	2.4 G: 3.56	NA
[3]	Monopole Loop	Eyewear	Wi-Fi 2.4 G Mono: 2.32–2.35 Loop: 2.35–2.5	Monopole: 10.7 Loop: 33.9	Monopole: 11.6 Loop: 2.7	2 × 2
[4]	Array	Eyewear	WiGig 57–66	WiGig: 58	60G: 11	NA
[14]	Patch	Eyewear	Wi-Fi 2.4 G 2.43–2.48	2.45 G: 23	2.45 G: −1.08	2 × 2
[20]	CE	Eyewear	LTE 0.7–2.7	LB: 58 HB: 63	LB: −5.03 HB: 0.65	NA
[28]	Array	Eyewear	60 G 59.05–60.1	NA	60 G: 9.29	NA
[29]	Loop Slot	Eyewear	LTE/Sub-6 G 0.82–0.96 1.71–2.69 3.3–3.6 4.8–5.0	LB: 86.5 HB: 98	LB: 2.67 HB: 3.72	LTE 2 × 2 Sub-6 G 4 × 4
[30]	Frame	Eyewear	IOT 5.8 G 5.42–6.27	NA	5.8 G: 8.18	NA
This work	Slot	Eyewear	Wi-Fi 5G/ Wi-Fi 6e 4.58–5.72 6.38–7.0	5.1 G: 85.7 6.8 G: 82.8	5.1 G: 4.3 6.8 G: 3.3	4 × 4

5. Conclusions

This paper proposes a method that allows standard plastic glasses in daily life to undergo a unique antenna manufacturing process, which can also be applied to Wi-Fi 5G and Wi-Fi 6e communication bands. Mainly relying on the antenna LDS process, four symmetrical slot antennas are carved on the plastic temples so that the plastic glasses can also achieve the 4 × 4 MIMO transmission effect. Furthermore, such an antenna configuration can make full use of the glasses' space to receive the signal on the left and right without blind spots. At the end of this study, a physical antenna was successfully produced, and the actual measurement results show that the efficiency and gain of the antenna are good, meaning that the smart plastic glasses have a high probability of being used in Wi-Fi 5G and Wi-Fi 6e application scenarios in the future.

Author Contributions: Conceptualization, M.-A.C.; methodology, M.-A.C.; software, M.-A.C., C.-W.H., C.-W.Y., and B.-R.C.; validation, C.-W.H. and C.-W.Y.; formal analysis, M.-A.C., C.-W.H., and C.-W.Y.; investigation, M.-A.C.; resources, M.-A.C.; writing—original draft preparation, M.-A.C. and C.-W.H.; writing—review and editing, M.-A.C.; visualization, M.-A.C.; supervision, M.-A.C.; project administration, M.-A.C.; funding acquisition, M.-A.C. All authors have read and agreed to the published version of the manuscript.

Funding: This work was supported by the Ministry of Science and Technology, Taiwan, under Grant Project MOST 109-2222-E-027-008.

Data Availability Statement: All data are included within the manuscript.

Conflicts of Interest: The authors declare no conflict of interest.

References

1. Hong, S.; Kang, S.H.; Kim, Y.; Jung, C.W. Transparent and Flexible Antenna for Wearable Glasses Applications. *IEEE Trans. Antennas Propag.* **2016**, *64*, 2797–2804. [CrossRef]
2. Faith, K.; Atef, Z.E. Smart Glasses Radiation Effects on a Human Head Model at Wi-Fi and 5G Cellular Frequencies. In Proceedings of the International ACES Communications, Beijing, China, 29 July–1 August 2018.
3. Cihangir, A.; Gianesello, F.; Luxey, C. Dual Antenna Concept with Complementary Radiation for Eyewear Applications. *IEEE Trans. Antennas Propag.* **2018**, *66*, 3056–3063. [CrossRef]
4. Bisognin, A.; Cihangir, A.; Luxey, C.; Jacquemod, G.; Pilard, R.; Gianesello, F.; Costa, J.R.; Fernandes, C.A.; Lima, E.B.; Panagamuwa, C.J.; et al. Ball Grid Array-Module with Integrated Shaped Lens for WiGig Applications in Eyewear Devices. *IEEE Trans. Antennas Propag.* **2017**, *65*, 6380–6394. [CrossRef]
5. Cihangir, A.; Panagamuwa, C.J.; Whittow, W.G.; Gianesello, F.; Luxey, C. Ultrabroadband Antenna with Robustness to Body Detuning for 4G Eyewear Devices. *IEEE Trans. Antennas Propag.* **2017**, *16*, 1225–1228. [CrossRef]
6. Nakano, H.; Yamauchi, J. Printed Slot and Wire Antennas: A Review. *Proc. IEEE* **2012**, *100*, 2158–2168. [CrossRef]
7. Chen, S.C.; Hsu, M.C. LTE MIMO Closed Slot Antenna System for Laptops with a Metal Cover. *IEEE Access* **2019**, *7*, 28973–28981. [CrossRef]
8. Barani, I.R.R.; Wong, K. Integrated Inverted-F and Open-Slot Antennas in the Metal-Framed Smartphone for 2×2 LTE LB and 4×4 LTE M/HB MIMO Operations. *IEEE Trans. Antennas Propag.* **2018**, *66*, 5004–5012. [CrossRef]
9. Cai, Q.; Li, Y.; Zhang, X.; Shen, W. Wideband MIMO Antenna Array Covering 3.3—7.1 GHz for 5G Metal-Rimmed Smartphone Applications. *IEEE Access* **2019**, *7*, 142070–142084. [CrossRef]
10. Liao, C.T.; Yang, Z.K.; Chen, H.M. Multiple Integrated Antennas for Wearable Fifth-Generation Communication and Internet of Things Applications. *IEEE Access* **2021**, *9*, 120328–120346. [CrossRef]
11. Yao, L.; Li, E.; Yan, J.; Shan, Z.; Ruan, X.; Shen, Z.; Ren, Y.; Yang, J. Miniaturization and Electromagnetic Reliability of Wearable Textile Antennas. *Electronics* **2021**, *10*, 994. [CrossRef]
12. Li, K.; Dong, T.; Xia, Z. Wideband Printed Wide-Slot Antenna with Fork-Shaped Stub. *Electronics* **2019**, *8*, 347. [CrossRef]
13. Zhang, S.; Zhao, K.; Ying, Z.; He, S. Adaptive Quad-Element Multi-Wideband Antenna Array for User-Effective LTE MIMO Mobile Terminals. *IEEE Trans. Antennas Propag.* **2013**, *61*, 4275–4283. [CrossRef]
14. Choi, S.; Choi, J. Miniaturized MIMO Antenna with a High Isolation for Smart Glasses. In Proceedings of the IEEE-APS Topical Conference on Antennas and Propagation in Wireless Communications, Verona, Italy, 11–15 September 2017.
15. Tang, Z.; Wu, X.; Zhan, J.; Hu, S.; Xi, Z.; Liu, Y. Compact UWB-MIMO Antenna with High Isolation and Triple Band-Notched Characteristics. *IEEE Access* **2019**, *7*, 19856–19865. [CrossRef]
16. Kulkarni, J.; Desai, A.; Desmond, C.Y. Wideband Four-Port MIMO Antenna Array with High Isolation for Future Wireless Systems. *Int. J. Electron. Commun.* **2021**, *128*, 1–15. [CrossRef]
17. Haq, M.A.U.; Koziel, S. Ground Plane Alterations for Design of High-Isolation Compact Wideband MIMO Antenna. *IEEE Access* **2018**, *6*, 48978–48983. [CrossRef]
18. Khalid, H.; Awan, W.A.; Hussain, M.; Fatima, A.; Ali, M.; Hussain, N.; Khan, S.; Alibakhshikenari, M.; Limiti, E. Design of an Integrated Sub-6 GHz and mmWave MIMO Antenna for 5G Handheld Devices. *Appl. Sci.* **2021**, *11*, 8331. [CrossRef]
19. Le, T.T.; Yun, T.Y. Wearable Dual-Band High-Gain Low-SAR Antenna for Off-Body Communication. *IEEE Antennas Wirel. Propag. Lett.* **2021**, *20*, 1175–1179. [CrossRef]
20. Cihangir, A.; Panagamuwa, C.J.; Whittow, W.G.; Jacquemod, G.; Gianesello, F.; Pilard, R.; Luxey, C. Dual-band 4G eyewear antenna and sar implications. *IEEE Trans. Antennas Propag.* **2017**, *65*, 2085–2089. [CrossRef]
21. RF Exposure Procedures and Equipment Authorization Policies for Mobile and Portable Devices. Available online: https://apps.fcc.gov/oetcf/kdb/forms/FTSSearchResultPage.cfm?switch=P&id=20676 (accessed on 23 October 2015).
22. IEEE Recommended Practice for Determining the Peak Spatial-Average Specific Absorption Rate (SAR) in the Human Head from Wireless Communications Devices: Measurement Techniques. Available online: https://ieeexplore.ieee.org/document/6589093 (accessed on 6 September 2013).
23. FCC Wi-Fi 6E RF Exposure—FCC Report. Available online: https://fcc.report/FCC-ID/MSQI005D/5126834.pdf (accessed on 4 February 2021).
24. Chung, M.A.; Chang, W.H. Low-cost, low-profile and miniaturized single-plane antenna design for an Internet of Thing device applications operating in 5G, 4G, V2X, DSRC, WiFi 6 band, WLAN, and WiMAX communication systems. *Microw. Opt. Technol. Lett.* **2020**, *62*, 1765–1773. [CrossRef]
25. Abdullah, M.; Kiani, S.H.; Abdulrazak, L.F.; Iqbal, A.; Bashir, M.A.; Khan, S.; Kim, S. High-Performance Multiple-Input Multiple-Output Antenna System for 5G Mobile Terminals. *Electronics* **2019**, *8*, 1090. [CrossRef]
26. Yang, P. Reconfigurable 3-D Slot Antenna Design for 4G and Sub-6G Smartphones with Metallic Casing. *Electronics* **2020**, *9*, 216. [CrossRef]
27. He, W.; Xu, B.; Yao, Y.; Colombi, D.; Ying, Z.; He, S. Implications of Incident Power Density Limits on Power and EIRP Levels of 5G Millimeter-Wave User Equipment. *IEEE Access* **2020**, *8*, 148214–148225. [CrossRef]
28. Hong, Y.; Chou, J. 60 GHz Patch Antenna Array with Parasitic Elements for Smart Glasses. *IEEE Antennas Wirel. Propag. Lett.* **2018**, *17*, 1252–1256. [CrossRef]

29. Wang, Y.Y.; Ban, Y.L.; Liu, Y. Sub-6GHz 4G/5G Conformal Glasses Antennas. *IEEE Access* **2019**, *7*, 182027–182036. [CrossRef]
30. Wang, Y.; Zhang, J.; Peng, F.; Wu, S. A Glasses Frame Antenna for the Applications in Internet of Things. *IEEE Internet Things J.* **2019**, *6*, 8911–8918. [CrossRef]

MDPI

St. Alban-Anlage 66

4052 Basel

Switzerland

Tel. +41 61 683 77 34

Fax +41 61 302 89 18

www.mdpi.com

Electronics Editorial Office

E-mail: electronics@mdpi.com

www.mdpi.com/journal/electronics